T0303883

Topics in Galois Theory

Research Notes in Mathematics

Volume 1

Topics in Galois Theory

Second Edition

Jean-Pierre Serre
Collège de France

Notes written by Henri Darmon
McGill University

A K Peters, Ltd.
Wellesley, Massachusetts

CRC Press
Taylor & Francis Group
6000 Broken Sound Parkway NW, Suite 300
Boca Raton, FL 33487-2742

© 2007 by Taylor & Francis Group, LLC
CRC Press is an imprint of Taylor & Francis Group, an Informa business

No claim to original U.S. Government works

Printed on acid-free paper
Version Date: 20140313

International Standard Book Number-13: 978-1-56881-412-4 (Hardback)

Visit the Taylor & Francis Web site at
http://www.taylorandfrancis.com

and the CRC Press Web site at
http://www.crcpress.com

Contents

Foreword

These notes are based on "Topics in Galois Theory," a course given by J-P. Serre at Harvard University in the Fall semester of 1988 and written down by H. Darmon. The course focused on the inverse problem of Galois theory: the construction of field extensions having a given finite group G as Galois group, typically over $\mathbf{Q}$ but also over fields such as $\mathbf{Q}(T)$.

Chapter 1 discusses examples for certain groups G of small order. The method of Scholz and Reichardt, which works over $\mathbf{Q}$ when G is a p-group of odd order, is given in chapter 2. Chapter 3 is devoted to the Hilbert irreducibility theorem and its connection with weak approximation and the large sieve inequality. Chapters 4 and 5 describe methods for showing that G is the Galois group of a regular extension of $\mathbf{Q}(T)$ (one then says that G has property Gal_T). Elementary constructions (e.g. when G is a symmetric or alternating group) are given in chapter 4, while the method of Shih, which works for $G = \mathbf{PSL}_2(p)$ in some cases, is outlined in chapter 5. Chapter 6 describes the GAGA principle and the relation between the topological and algebraic fundamental groups of complex curves. Chapters 7 and 8 are devoted to the rationality and rigidity criterions and their application to proving the property Gal_T for certain groups (notably, many of the sporadic simple groups, including the Fischer-Griess Monster). The relation between the Hasse-Witt invariant of the quadratic form $\mathrm{Tr}\,(x^2)$ and certain embedding problems is the topic of chapter 9, and an application to showing that $\tilde{A}_n$ has property Gal_T is given. An appendix (chapter 10) gives a proof of the large sieve inequality used in chapter 3.

The reader should be warned that most proofs only give the main ideas; details have been left out. Moreover, a number of relevant topics have been omitted, for lack of time (and understanding), namely:

a) The theory of generic extensions, cf. [Sa1].

b) Shafarevich's theorem on the existence of extensions of $\mathbf{Q}$ with a given solvable Galois group, cf. [NSW], chap. IX.

c) The Hurwitz schemes which parametrize extensions with a given Galois group and a given ramification structure, cf. [Fr1], [Fr2], [Ma3].

d) The computation of explicit equations for extensions with Galois group $\mathbf{PSL}_2(\mathbf{F}_7)$, $\mathbf{SL}_2(\mathbf{F}_8)$, M_{11}, ..., cf. [LM], [Ma3], [Ma4], [Ml1], ...

e) Mestre's results [Me3] on extensions of $\mathbf{Q}(T)$ with Galois group $6 \cdot A_6$, $6 \cdot A_7$, and $\mathbf{SL}_2(\mathbf{F}_7)$.

We wish to thank Larry Washington for his helpful comments on an earlier version of these notes.

Paris, August 1991.

H. Darmon   J-P. Serre

For the second edition of these Notes, some corrections have been made, and the references have been updated.

Paris, June 2004

J-P. Serre

Notation

If V is an algebraic variety over the field K, and L is an extension of K, we denote by $V(L)$ the set of L-points of V and by $V_{/L}$ the L-variety obtained from V by base change from K to L. All the varieties are supposed reduced and quasi-projective.

$\mathbf{A}^n$ is the affine n-space; $\mathbf{A}^n(L) = L^n$.

$\mathbf{P}_n$ is the projective n-space; $\mathbf{P}_n(L) = (L^{(n+1)} - \{0\})/L^*$; the group of automorphisms of $\mathbf{P}_n$ is $\mathbf{PGL}_n = \mathbf{GL}_n/\mathbf{G}_m$.

If X is a finite set, $|X|$ denotes the cardinality of X.

Introduction

The question of whether all finite groups can occur as Galois groups of an extension of the rationals (known as the *inverse problem* of Galois theory) is still unsolved, in spite of substantial progress in recent years.

In the 1930's, Emmy Noether proposed the following strategy to attack the inverse problem [Noe]: by embedding G in S_n, the permutation group on n letters, one defines a G-action on the field $\mathbf{Q}(X_1, \ldots, X_n) = \mathbf{Q}(\underline{X})$. Let E be the fixed field under this action. Then $\mathbf{Q}(\underline{X})$ is a Galois extension of E with Galois group G.

In geometric terms, the extension $\mathbf{Q}(\underline{X})$ of E corresponds to the projection of varieties: $\pi : \mathbf{A}^n \longrightarrow \mathbf{A}^n/G$, where $\mathbf{A}^n$ is affine n-space over $\mathbf{Q}$. Let P be a $\mathbf{Q}$-rational point of $\mathbf{A}^n/G$ for which π is unramified, and lift it to $Q \in \mathbf{A}^n(\bar{\mathbf{Q}})$. The conjugates of Q under the action of $\mathrm{Gal}(\bar{\mathbf{Q}}/\mathbf{Q})$ are the sQ where $s \in H_Q \subset G$, and H_Q is the decomposition group at Q. If $H_Q = G$, then Q generates a field extension of $\mathbf{Q}$ with Galois group G.

A variety is said to be *rational over* $\mathbf{Q}$ (or $\mathbf{Q}$-*rational*) if it is birationally isomorphic over $\mathbf{Q}$ to the affine space $\mathbf{A}^n$ for some n, or equivalently, if its function field is isomorphic to $\mathbf{Q}(T_1, \ldots, T_n)$, where the T_i are indeterminates.

Theorem 1 (Hilbert, [Hi]) *If $\mathbf{A}^n/G$ is $\mathbf{Q}$-rational, then there are infinitely many points P, Q as above such that $H_Q = G$.*

This follows from Hilbert's irreducibility theorem, cf. §3.4.

Example: Let $G = S_n$, acting on $\mathbf{Q}(X_1, \ldots, X_n)$. The field E of S_n-invariants is $\mathbf{Q}(T_1, \ldots, T_n)$, where T_i is the ith symmetric polynomial, and $\mathbf{Q}(X_1, \ldots, X_n)$ has Galois group S_n over E: it is the splitting field of the polynomial

$$X^n - T_1 X^{n-1} + T_2 X^{n-2} + \cdots + (-1)^n T_n.$$

Hilbert's irreducibility theorem says that the T_i can be specialized to infinitely many values $t_i \in \mathbf{Q}$ (or even $t_i \in \mathbf{Z}$) such that the equation

$$X^n - t_1 X^{n-1} + t_2 X^{n-2} + \cdots + (-1)^n t_n = 0$$

has Galois group S_n over $\mathbf{Q}$. In fact, "most" t_i work: of the N^n n-tuples (t_i) with $t_i \in \mathbf{Z}$, $1 \leq t_i \leq N$, only $O(N^{n-\frac{1}{2}} \log N)$ may fail to give S_n, cf. [Ga], [Coh], [Se9].

In addition to the symmetric groups, the method works for the alternating groups A_n with $n \leq 5$, cf. [Mae] (For $n \geq 6$, it is not known whether the field of A_n-invariants is rational.) Somewhat surprisingly, there are groups for which the method fails (i.e. $\mathbf{A}^n/G$ is not $\mathbf{Q}$-rational):

• Swan [Sw1] has shown that the field of G-invariants is not rational when G is a cyclic group of order 47. The obstruction is related to the automorphism group of G which is a cyclic group of order $46 = 2 \times 23$, and to the fact that $\mathbf{Q}(\zeta_{23})$ does not have class number 1 (since $h(-23) = 3$).

• In [Le] H. Lenstra gives a general criterion for the field of G-invariants to be rational when G is an abelian group: in particular, he shows that this criterion is not satisfied when G is cyclic of order 8.

(The above counter-examples are over $\mathbf{Q}$. Counter-examples over $\mathbf{C}$ (involving a non-abelian group G) are given by the following result of Saltman [Sa2]: if there is a non-zero $\alpha \in H^2(G, \mathbf{Q}/\mathbf{Z})$ such that $\mathrm{Res}_H^G(\alpha) = 0$ for all abelian subgroups H generated by two elements, then $\mathbf{A}^n/G$ is not $\mathbf{C}$-rational. It is not hard to construct groups G satisfying the hypothesis of Saltman's theorem: for example, one may take a suitable extension of abelian groups of type $(p, \dots, p)$.)

It is easy to see (e.g., using the normal basis theorem) that the covering map

$$\pi : \mathbf{A}^n \longrightarrow \mathbf{A}^n/G$$

is *generic* (or *versal*) in the sense that every extension of $\mathbf{Q}$ (or of any field of characteristic zero) with Galois group G can be obtained by taking the π-fibre of a rational point of $\mathbf{A}^n/G$ over which π is unramified. Hence, if $\mathbf{A}^n/G$ is $\mathbf{Q}$-rational, then the set of all G-extensions of $\mathbf{Q}$ can be described by a system of n rational parameters. Such a parametrization implies the following property of extensions with Galois group G [Sa1]:

Theorem 2 *Assume $\mathbf{A}^n/G$ is $\mathbf{Q}$-rational. Let $\{p_i\}$ be a finite set of primes, L_i extensions of $\mathbf{Q}_{p_i}$ with Galois group G. Then there is an extension L of $\mathbf{Q}$ with $\mathrm{Gal}(L/\mathbf{Q}) = G$ such that $L \otimes \mathbf{Q}_{p_i} = L_i$.*

Remark: There is a more general statement, where the L_i are allowed to be Galois algebras, and $\mathbf{Q}$ is replaced by a field endowed with finitely many independent absolute values.

Proof (sketch): Each L_i is parametrized by $(\underline{X}^{(i)}) \in \mathbf{A}^n(\mathbf{Q}_{p_i})$. A "global" parameter $(\underline{X}) \in \mathbf{A}^n(\mathbf{Q})$ which is sufficiently close to each of the $(\underline{X}^{(i)})$ in the $\mathbf{Q}_{p_i}$-topology gives an extension of $\mathbf{Q}$ with group G having the desired local behaviour (Krasner's lemma). QED.

The cyclic group of order 8 does not satisfy the property of th. 2. Indeed, if $L_2/\mathbf{Q}_2$ is the unique unramified extension of $\mathbf{Q}_2$ of degree 8, there is no cyclic extension L of degree 8 over $\mathbf{Q}$ such that $L_2 \simeq L \otimes \mathbf{Q}_2$ (an easy exercise on characters, see [Wa]).

One could perhaps extend Hilbert's theorem to a more general class of varieties. There is an interesting suggestion of Ekedahl and Colliot-Thélène in this direction [Ek], [CT] (see §3.5).

Since A^n/G is not always $\mathbf{Q}$-rational, one has to settle for less:

Question: *If G is a finite group, can it be realized as a Galois group of some $\mathbf{Q}$-regular extension F of $\mathbf{Q}(T)$?* (Recall that "F is $\mathbf{Q}$-regular" means that $F \cap \bar{\mathbf{Q}} = \mathbf{Q}$; in what follows we shall usually write "regular" instead of "$\mathbf{Q}$-regular".)

Remarks:
1. If F is a function field of a variety V defined over $\mathbf{Q}$, then F is regular if and only if V is absolutely irreducible. The regularity assumption is included to rule out uninteresting examples such as the extension $E(T)$ of $\mathbf{Q}(T)$ where E is a Galois extension of $\mathbf{Q}$.
2. If such an F exists, then there are infinitely many linearly disjoint extensions of $\mathbf{Q}$ with Galois group G.

The existence of regular extensions of $\mathbf{Q}(T)$ with Galois group G is known when G is:
• Abelian;
• One of the 26 sporadic simple groups, (with the possible exception of the Mathieu group M_{23});
•$\mathbf{PSL}_2(\mathbf{F}_p)$, where at least one of $\left(\frac{2}{p}\right), \left(\frac{3}{p}\right), \left(\frac{7}{p}\right)$ is -1 [Shih1], [Shih2];
• A_n, or S_n, cf. [Hi];
• $\tilde{A}_n$, cf. N. Vila [Vi] and J-F. Mestre [Me2];
• $G_2(\mathbf{F}_p)$ [Th2], where G_2 is the automorphism group of the octonions, and the list is not exhaustive.

The method for finding F proceeds as follows:

1. Construction (by analytic and topological methods) of an extension $F_{\mathbf{C}}$ of $\mathbf{C}(T)$ with Galois group G.

2. A descent from $\mathbf{C}$ to $\mathbf{Q}$. This is the hardest part, and requires that G satisfies a so-called *rigidity* criterion.

The outline of the course will be:

1. Elementary examples, and the Scholz-Reichardt theorem.

2. Hilbert's irreducibility theorem and applications.

3. The "rigidity method" used to obtain extensions of $\mathbf{Q}(T)$ with given Galois groups.

4. The quadratic form $x \mapsto \mathrm{Tr}\,(x^2)$, and its applications to embedding problems, e.g., construction of extensions with Galois group $\tilde{A}_n$.

Chapter 1

Examples in low degree

1.1   The groups $\mathbf{Z}/2\mathbf{Z}$, $\mathbf{Z}/3\mathbf{Z}$, and S_3

- $G = \mathbf{Z}/2\mathbf{Z}$: all quadratic extensions can be obtained by taking square roots: the map $\mathbf{P}_1 \longrightarrow \mathbf{P}_1$ given by $X \mapsto X^2$ is generic (in characteristic different from 2 - for a characteristic-free equation, one should use $X^2 - TX + 1 = 0$ instead).

- $G = \mathbf{Z}/3\mathbf{Z}$: A "generic equation" for G is:

$$X^3 - TX^2 + (T - 3)X + 1 = 0,$$

with discriminant $\Delta = (T^2 - 3T + 9)^2$. (In characteristic 3, this reduces to the Artin-Schreier equation $Y^3 - Y = -1/T$ by putting $Y = 1/(X + 1)$.) The group G acts on $\mathbf{P}_1$ by

$$\sigma X = \frac{1}{1 - X},$$

where σ is a generator of G. The function

$$T = X + \sigma X + \sigma^2 X = \frac{X^3 - 3X + 1}{X^2 - X}$$

is G-invariant and gives a map $Y = \mathbf{P}_1 \longrightarrow \mathbf{P}_1/G$. To check genericity, observe that any extension L/K with cyclic Galois group of order 3 defines a homomorphism $\phi : G_K \longrightarrow G \longrightarrow \operatorname{Aut} Y$ which can be viewed as a 1-cocycle with values in $\operatorname{Aut} Y$. The extension L/K is given by a rational point on $\mathbf{P}_1/G$ if and only if the twist of Y by this cocycle has a rational point not invariant by σ. This is a general property of Galois twists. But this twist has a rational point over a cubic extension of K, and every curve of genus 0 which has a point over an odd-degree extension is a projective line, and hence has at least one rational point distinct from the ones fixed by σ.

- $G = S_3$: The map

$$S_3 \hookrightarrow \mathbf{GL}_2 \longrightarrow \mathbf{PGL}_2 = \mathrm{Aut}\,(\mathbf{P}_1)$$

gives a projection

$$\mathbf{P}_1 \longrightarrow \mathbf{P}_1/S_3 = \mathbf{P}_1$$

which is generic, although the reasoning for C_3 cannot be applied, as the order of S_3 is even. But S_3 can be lifted from $\mathbf{PGL}_2$ to $\mathbf{GL}_2$, and the vanishing of $H^1(G_K, \mathbf{GL}_2)$ can be used to show that

$$\mathbf{P}_1 \longrightarrow \mathbf{P}_1/S_3$$

is generic.

Exercise: Using the above construction (or a direct argument), show that every separable cubic extension of K is given by an equation of the form

$$X^3 + TX + T = 0, \quad \text{with } T \neq 0, -27/4.$$

1.2 The group C_4

Let K_4/K be Galois and cyclic of degree 4, and suppose that Char $K \neq 2$. The extension K_4 is obtained from a unique tower of quadratic extensions:

$$K \subset K_2 \subset K_4,$$

where $K_2 = K(\sqrt{\epsilon})$, and $K_4 = K_2(\sqrt{a + b\sqrt{\epsilon}})$.

Conversely, let $K_2 = K(\sqrt{\epsilon})$ be a quadratic extension of K, where $\epsilon \in K^*$ is not a square. If $a, b \in K$ and $K_4 = K_2(\sqrt{a + b\sqrt{\epsilon}})$, then K_4 may not be Galois over K (its Galois closure could have Galois group isomorphic to D_4, the dihedral group of order 8).

Theorem 1.2.1 *The field K_4 is cyclic of degree 4 if and only if $a^2 - \epsilon b^2 = \epsilon c^2$ for some $c \in K^*$.*

Proof: Let G be a group, ϵ a non-trivial homomorphism from G to $\mathbf{Z}/2\mathbf{Z}$, and χ a homomorphism from $H = \mathrm{Ker}\,\epsilon$ to $\mathbf{Z}/2\mathbf{Z}$. Let H_χ denote the kernel of χ.

Lemma 1.2.2 *The following are equivalent*:

(a) H_χ *is normal in* G, *and* G/H_χ *is cyclic of order* 4.

(b) $\mathrm{Cor}_H^G\chi = \epsilon$, *where* Cor_H^G *is the corestriction map.*

(We abbreviate $H^1(G, \mathbf{Z}/2\mathbf{Z}) = \mathrm{Hom}(G, \mathbf{Z}/2\mathbf{Z})$ to $H^1(G)$. The corestriction map $H^1(H) \longrightarrow H^1(G)$ can be defined by

$$(\mathrm{Cor}_H^G\chi)(g) = \chi(\mathrm{Ver}_H^G g),$$

where $\mathrm{Ver}_H^G : G/(G, G) \longrightarrow H/(H, H)$ is the *transfer*.)

The proof that (a) $\Rightarrow$ (b) is immediate: replacing G by G/H_χ, it suffices to check that the transfer $C_4 \longrightarrow C_2$ is onto: but this map is given by $s \longmapsto s^2$.

Now, assume (b). Select $s \in G - H$. The transfer is given by:

$$\mathrm{Ver}_H^G(h) = h \cdot shs^{-1} \bmod (H, H).$$

Hence for all $h \in H$:

$$\chi(\mathrm{Ver}_H^G h) = \chi(h) + \chi(shs^{-1}) = \epsilon(h) \equiv 0 \pmod 2.$$

But if $h \in H_\chi$, then $\chi(h) = 0$. It follows that $\chi(shs^{-1}) = 0$, so that H_χ is normal in G. Now, applying the hypothesis to s shows

$$\chi(s^2) = \mathrm{Cor}_H^G\chi(s) = \epsilon(s) \equiv 1 \pmod 2,$$

so $s^2 \neq 1 \pmod{H_\chi}$. It follows that G/H_χ is cyclic of order 4, and this completes the proof of lemma 1.2.2.

Now, let $G = G_K = \mathrm{Gal}(\bar{K}/K)$. The extensions K_2 and K_4 define homomorphisms ϵ and χ as in the lemma. Via the identification of $H^1(G_K)$ with K^*/K^{*2}, the corestriction map $\mathrm{Cor} : H^1(G_{K_2}) \longrightarrow H^1(G_K)$ is equal to the norm, and the criterion $\mathrm{Cor}_H^G\chi = \epsilon$ becomes:

$$N(a + b\sqrt{\epsilon}) = \epsilon c^2,$$

where $c \in K^*$. This completes the proof of th. 1.2.1.

Remark: In characteristic 2, Artin-Schreier theory gives an isomorphism $H^1(G_K) \simeq K/\wp K$, where $\wp x = x^2 + x$, and the corestriction map corresponds to the trace. Hence the analogue of th. 1.2.1 in characteristic 2 is:

Theorem 1.2.3 *Suppose* $\mathrm{Char}\, K = 2$, *and let* $K_2 = K(x), K_4 = K_2(y)$, *where* $\wp x = \epsilon, \wp y = a + bx$. *Then* K_4 *is Galois over* K *and cyclic of degree* 4 *if and only if* $\mathrm{Tr}\,(a + bx)(= b)$ *is of the form* $\epsilon + z^2 + z$, *with* $z \in K$.

Observe that the variables ϵ, a, z of th. 1.2.3 parametrize C_4-extensions of K. In particular, it is possible in characteristic 2 to embed any quadratic extension in a cyclic extension of degree 4. This is a special case of a general result: the embedding problem for p-groups always has a solution in characteristic p (as can be seen from the triviality of $H^2(G, P)$ when G is the absolute Galois group of a field of characteristic p and P is an abelian p-group with G-action. See for example [Se1].)

The situation is different in characteristic $\neq 2$: the criterion $a^2 - b^2\epsilon = \epsilon c^2$ implies that ϵ must be a sum of 2 squares in K: if $b^2 + c^2 \neq 0$, then:

$$\epsilon = \left(\frac{ab}{b^2 + c^2}\right)^2 + \left(\frac{ac}{b^2 + c^2}\right)^2.$$

Otherwise $\sqrt{-1} \in K$, and any element of K can be expressed as a sum of 2 squares. Conversely, if ϵ is the sum of two squares, $\epsilon = \lambda^2 + \mu^2$, then setting

$$a = \lambda^2 + \mu^2, \quad b = \lambda, \quad c = \mu,$$

solves the equation $a^2 - b^2\epsilon = c^2\epsilon$. Hence we have shown:

Theorem 1.2.4 *A quadratic extension $K(\sqrt{\epsilon})$ can be embedded in a cyclic extension of degree 4 if and only if ϵ is a sum of two squares in K.*

Here is an alternate proof of th. 1.2.4: the quadratic extension K_2 can be embedded in a cyclic extension K_4 of degree 4 if and only if the homomorphism $\epsilon : G_K \longrightarrow \mathbf{Z}/2\mathbf{Z}$ given by K_2 factors through a homomorphism $G_K \longrightarrow \mathbf{Z}/4\mathbf{Z}$. This suggests that one apply Galois cohomology to the sequence:

$$0 \longrightarrow \mathbf{Z}/2\mathbf{Z} \longrightarrow \mathbf{Z}/4\mathbf{Z} \longrightarrow \mathbf{Z}/2\mathbf{Z} \longrightarrow 0,$$

obtaining:

$$H^1(G_K, \mathbf{Z}/4\mathbf{Z}) \longrightarrow H^1(G_K, \mathbf{Z}/2\mathbf{Z}) \xrightarrow{\delta} H^2(G_K, \mathbf{Z}/2\mathbf{Z}).$$

The obstruction to lifting $\epsilon \in H^1(G_K, \mathbf{Z}/2\mathbf{Z})$ to $H^1(G_K, \mathbf{Z}/4\mathbf{Z})$ is given by $\delta\epsilon \in H^2(G_K, \mathbf{Z}/2\mathbf{Z}) = \mathrm{Br}_2(K)$, where $\mathrm{Br}_2(K)$ denotes the 2-torsion in the Brauer group of K. It is well-known that the connecting homomorphism $\delta : H^1 \longrightarrow H^2$, also known as the Bockstein map, is given by $\delta x = x \cdot x$ (cup-product). This can be proved by computing on the "universal example" $\mathbf{P}_\infty(\mathbf{R}) = K(\mathbf{Z}/2\mathbf{Z}, 1)$ which is the classifying space for $\mathbf{Z}/2\mathbf{Z}$. The cup product can be computed by the formula:

$$\alpha \cdot \beta = (\alpha, \beta),$$

where $H^1(G, \mathbf{Z}/2\mathbf{Z})$ is identified with K^*/K^{*2} and (α, β) denotes the class of the quaternion algebra given by

$$i^2 = \alpha, j^2 = \beta, ij = -ji.$$

But $(\epsilon, -\epsilon) = 0$ (in additive notation), so $(\epsilon, \epsilon) = (-1, \epsilon)$. Hence, $\delta\epsilon$ is 0 if and only if $(-1, \epsilon) = 0$, i.e., ϵ is a sum of two squares in K.

Similarly, one could ask when the extension K_4 can be embedded in a cyclic extension of degree 8. The obstruction is again given by an element of $\mathrm{Br}_2(K)$. One can prove (e.g., by using [Se6]):

Theorem 1.2.5 *The obstruction to embedding the cyclic extension K_4 in a cyclic extension of degree 8 is given by the class of $(2, \epsilon) + (-1, a)$ in $\mathrm{Br}_2(K)$, if $a \neq 0$, and by the class of $(2, \epsilon)$ if $a = 0$.*

Hence, when $a \neq 0$, the C_8-embedding problem is possible if and only if the quaternion algebra $(2, \epsilon)$ is isomorphic to $(-1, a)$.

There is also a direct proof of th. 1.2.1. Let K_2 and K_4 be as before, with $K_2 = K(\sqrt{\epsilon})$, $K_4 = K_2(\sqrt{a + b\sqrt{\epsilon}})$. Let $x = \sqrt{a + b\sqrt{\epsilon}}$ and $y = \sqrt{a - b\sqrt{\epsilon}}$. If $\mathrm{Gal}(K_4/K) = C_4$, then we may choose a generator σ of $\mathrm{Gal}(K_4/K)$ taking x to y, and hence y to $-x$. Setting

$$c = xy/\sqrt{\epsilon},$$

we have $\sigma c = y(-x)/(-\sqrt{\epsilon}) = c$, so $c \in K^*$. Also

$$\epsilon c^2 = x^2 y^2 = a^2 - b^2\epsilon,$$

and one obtains the same criterion as before. Conversely, if a, b, c, ϵ satisfy the equation $a^2 - b^2\epsilon = c^2\epsilon$, one verifies that K_4 is a cyclic extension with Galois group C_4.

Remarks:
1. The minimal polynomial for x over K is

$$X^4 + AX^2 + B = 0,$$

where $A = 2a$, $B = a^2 - \epsilon b^2$. The condition for a general polynomial of this form to have Galois group C_4 is that $A^2 - 4B$ is not a square and that

$$\left(\frac{A^2 - 4B}{B}\right) \in K^{*2}.$$

2. The C_4-extensions are parametrized by the solutions (ϵ, a, b, u) of the equation

$$a^2 - \epsilon b^2 = \epsilon u^2,$$

with $u \neq 0$ and ϵ not a square. This represents a rational variety: one can solve for ϵ in terms of a, b, and u. Hence the class of C_4-extensions of $\mathbf{Q}$ satisfies the conclusion of th. 2: there are C_4-extensions of $\mathbf{Q}$ with arbitrarily prescribed local behaviour at finitely many places; recall that this is not true for the cyclic group of order 8.

Exercises:
1. The group C_4 acts faithfully on $\mathbf{P}_1$ via the map $C_4 \longrightarrow \mathbf{PGL}_2$ which sends a generator σ of C_4 to $\begin{pmatrix} 1 & 1 \\ -1 & 1 \end{pmatrix}$. The corresponding map $\mathbf{P}_1 \longrightarrow \mathbf{P}_1/C_4$ is given by $z \mapsto (z^4 - 6z^2 + 1)/(z(z^2 - 1))$. This gives rise to the equation:

$$Z^4 - TZ^3 - 6Z^2 + TZ + 1 = 0$$

with Galois group C_4 over $\mathbf{Q}(T)$.
1.1 If $i \in K$, show that this equation is generic: in fact, it is equivalent to the Kummer equation.
1.2 If $i \notin K$, show that there does not exist any one-dimensional generic family for C_4-extensions.
1.3 If a C_4-extension is described as before by parameters ϵ, a, b and c, show that it comes from the equation above if and only if $(-1, a) = 0$ or $a = 0$.

2. Assume K contains a primitive 2^n-th root of unity z. Let $L = K(\sqrt[2^n]{a})$ be a cyclic extension of K of degree 2^n. Show that the obstruction to the embedding of L in a cyclic extension of degree 2^{n+1} is (a, z) in $\mathrm{Br}_2(K)$.

1.3   Application of tori to abelian Galois groups of exponent $2, 3, 4, 6$

A K-torus is an algebraic group over K which becomes isomorphic to a product of multiplicative groups $\mathbf{G}_m \times \ldots \times \mathbf{G}_m$ over the algebraic closure $\bar{K}$ of K. If this isomorphism is defined over K, then the torus is said to be *split*.

Let T be a K-torus and denote by $X(T)$ its *character group*,

$$X(T) = \mathrm{Hom}_{\bar{K}}(T, \mathbf{G}_m).$$

It is well known that $X(T)$ is a free $\mathbf{Z}$-module of rank $n = \dim T$ endowed with the natural action of G_K. The functor $T \mapsto X(T)$ defines an anti-equivalence between the category of finite dimensional tori over K and the category of free $\mathbf{Z}$-modules of finite rank with G_K action.

A split K-torus is clearly a K-rational variety; the same holds for tori which split over a quadratic extension K' of K. This follows from the classification of tori which split over a quadratic extension (whose proof we shall omit - see [CR]):

Lemma 1.3.1 *A free* $\mathbf{Z}$*-module of finite rank with an action of* $\mathbf{Z}/2\mathbf{Z}$ *is a direct sum of indecomposable modules of the form:*
1. $\mathbf{Z}$ *with trivial action.*
2. $\mathbf{Z}$ *with the non-trivial action.*
3. $\mathbf{Z} \times \mathbf{Z}$ *with the "regular representation" of* $\mathbf{Z}/2\mathbf{Z}$ *which interchanges the two factors.*

The corresponding tori are:
1. $\mathbf{G}_{\mathrm{m}}$
2. *A "twisted form" of* $\mathbf{G}_{\mathrm{m}}$, *which corresponds to elements of norm 1 in (3) below.*
3. *The algebraic group* $R_{K'/K}\mathbf{G}_{\mathrm{m}}$ *obtained from* $\mathbf{G}_{\mathrm{m}/K'}$ *by "restriction of scalars" to* K (*cf.* §3.2.1).

It is not difficult to show that the three cases give rise to K-rational varieties, and the result follows.

If G is a finite group, the group $\mathcal{G}$ of invertible elements of the group algebra $\Lambda = K[G]$ defines an algebraic group over K. In characteristic 0, we have

$$\mathcal{G} \simeq \prod \mathbf{GL}_{n_i} \quad \text{over } \bar{K},$$

where the product is taken over all irreducible representations of G and the n_i denote the dimensions of these representations.

In particular, if G is commutative, then $\mathcal{G}$ is a torus with character group $\mathbf{Z}[\hat{G}]$, where $\hat{G} = \mathrm{Hom}_{\bar{K}}(G, \mathbf{G}_{\mathrm{m}})$. Therefore, $\mathcal{G}$ splits over the field generated by the values of the characters of G. There is an exact sequence of algebraic groups:

$$1 \longrightarrow G \longrightarrow \mathcal{G} \longrightarrow \mathcal{G}/G \longrightarrow 1,$$

and the covering map $\mathcal{G} \longrightarrow \mathcal{G}/G$ is generic for extensions of K with Galois group G. If G is of exponent $2, 3$, 4 or 6, then $\mathcal{G}$ splits over a quadratic extension, since the characters values lie in $\mathbf{Q}$, $\mathbf{Q}(\sqrt{3})$, or $\mathbf{Q}(i)$. By the previous result, $\mathcal{G}$ - and hence *a fortiori* $\mathcal{G}/G$ - is $\mathbf{Q}$-rational. So the abelian groups of exponent 2,3,4 or 6 yield to Noether's method (but not those of exponent 8).

Exercise: Show that all tori decomposed by a cyclic extension of degree 4 are rational varieties, by making a list of indecomposable integer representations of the cyclic group of order 4 (there are nine of these, of degrees $1, 1, 2, 2, 3, 3, 4, 4, 4$). See [Vo2].

Chapter 2

Nilpotent and solvable groups as Galois groups over Q

2.1 A theorem of Scholz-Reichardt

Our goal will be to prove the following theorem which is due to Scholz and Reichardt [Re]:

Theorem 2.1.1 *Every l-group, $l \neq 2$, can be realized as a Galois group over* **Q**. (Equivalently, every finite nilpotent group of odd order is a Galois group over **Q**.)

Remarks:
1. This is a special case of a theorem of Shafarevich: every solvable group can be realized as a Galois group over **Q**. [The proofs of that theorem given in [Sha1] and [Is] are known to contain a mistake relative to the prime 2 (see [Sha3]). In the notes appended to his Collected Papers, p.752, Shafarevich sketches a method to correct this. For a complete proof, see [NSW], ch.IX, §5.]
2. The proof yields somewhat more than the statement of the theorem. For example, if $|G| = l^N$, then the extension of **Q** with Galois group G can be chosen to be ramified at at most N primes. It also follows from the proof that any separable pro-l-group of finite exponent is a Galois group over **Q**.
3. The proof does not work for $l = 2$. It would be interesting to see if there is a way of adapting it to this case.
4. It is not known whether there is a regular Galois extension of $\mathbf{Q}(T)$ with Galois group G for an arbitrary l-group G.

An l-group can be built up from a series of central extensions by groups of order l. The natural approach to the problem of realizing an l-group G as a Galois group over $\mathbf{Q}$ is to construct a tower of extensions of degree l which ultimately give the desired G-extension. When carried out naively, this approach does not work, because the embedding problem cannot always be solved. The idea of Scholz and Reichardt is to introduce more stringent conditions on the extensions which are made at each stage, ensuring that the embedding problem has a positive answer.

Let $K/\mathbf{Q}$ be an extension with Galois group G, where G is an l-group. Choose $N \geq 1$ such that l^N is a multiple of the exponent of G, i.e., $s^{l^N} = 1$ for all $s \in G$. The property introduced by Scholz is the following:

Definition 2.1.2 *The extension $L/\mathbf{Q}$ is said to have property (S_N) if every prime p which is ramified in $L/\mathbf{Q}$ satisfies:*
1. $p \equiv 1 \pmod{l^N}$.
2. If v is a place of L dividing p, the inertia group I_v at v is equal to the decomposition group D_v.

Condition 2 is equivalent to saying that the local extension $L_v/\mathbf{Q}_p$ is totally ramified, or that its residue field is $\mathbf{F}_p$.

Now, let
$$1 \longrightarrow C_l \longrightarrow \tilde{G} \longrightarrow G \longrightarrow 1$$

be an exact sequence of l-groups with C_l central, cyclic of order l. The "embedding problem" for $\tilde{G}$ is to find a Galois extension $\tilde{L}$ of K containing L, with isomorphisms $\mathrm{Gal}(\tilde{L}/L) \simeq C_l$ and $\mathrm{Gal}(\tilde{L}/K) \simeq \tilde{G}$ such that the diagram

$$
\begin{array}{ccccccccc}
1 & \longrightarrow & C_l & \longrightarrow & \tilde{G} & \longrightarrow & G & \longrightarrow & 1 \\
 & & \| & & \| & & \| & & \\
1 & \longrightarrow & \mathrm{Gal}(\tilde{L}/L) & \longrightarrow & \mathrm{Gal}(\tilde{L}/K) & \longrightarrow & \mathrm{Gal}(L/K) & \longrightarrow & 1
\end{array}
$$

is commutative.

The Scholz-Reichardt theorem is a consequence of the following (applied inductively):

Theorem 2.1.3 *Let $L/\mathbf{Q}$ be Galois with Galois group G, and assume that L has property (S_N). Assume further that l^N is a multiple of the exponent of $\tilde{G}$. Then the embedding problem for L and $\tilde{G}$ has a solution $\tilde{L}$, which satisfies (S_N) and is ramified at at most one more prime than L. (Furthermore, one can require that this prime be taken from any set of prime numbers of density one.)*

The proof of th. 2.1.3 will be divided into two parts: first, for split extensions, then for non-split ones.

First part: the case $\tilde{G} \simeq G \times C_l$

Let $(p_1, \ldots, p_m)$ be the prime numbers ramified in L. Select a prime number q with the following properties:

1. $q \equiv 1 \pmod{l^N}$,
2. q splits completely in the extension $L/\mathbf{Q}$,
3. Every prime p_i, $(1 \le i \le m)$ is an l-th power in $\mathbf{F}_q$.

Taken together, these conditions mean that the prime q splits completely in the field $L(\sqrt[l^N]{1}, \sqrt[l]{p_1}, \ldots, \sqrt[l]{p_m})$. The following well-known lemma guarantees the existence of such a q:

Lemma 2.1.4 *If $E/\mathbf{Q}$ is a finite extension of $\mathbf{Q}$, then there are infinitely many primes which split completely in E. In fact, every set of density one contains such a prime.*

Proof: The second statement in the lemma is a consequence of Chebotarev's density theorem; the first part can be proved by a direct argument, without invoking Chebotarev. Assume E is Galois, and let f be a minimal polynomial with integral coefficients of a primitive element of E. Suppose there are only finitely many primes p_i which split completely in E or are ramified. Then $f(x)$ is of the form $\pm p_1^{m_1} \ldots p_k^{m_k}$, for $x \in \mathbf{Z}$. When x is between 1 and X, the number of distinct values taken by $f(x)$ is at least $\frac{1}{n} X$. But the number of values of $f(x)$ which can be written in the form $\pm p_1^{m_1} \ldots p_k^{m_k}$ is bounded by a power of $\log X$. This yields a contradiction.

Having chosen a q which satisfies the conditions above, fix a surjective homomorphism

$$\lambda : (\mathbf{Z}/q\mathbf{Z})^* \longrightarrow C_l.$$

(Such a λ exists because $q \equiv 1 \pmod{l}$.) We view λ as a Galois character. This defines a C_l-extension M_λ of $\mathbf{Q}$ which is ramified only at q, and is linearly disjoint from L. The compositum LM_λ therefore has Galois group $\tilde{G} = G \times C_l$. Let us check that LM_λ satisfies property (S_N). By our choice of q, we have $q \equiv 1 \pmod{l^N}$. It remains to show that $I_v = D_v$ at all ramified primes. If p is ramified in $L/\mathbf{Q}$, it splits completely in M_λ, and hence $D_v = I_v$ for all primes $v|p$. The only prime ramified in M_λ is q, and q splits completely in L by assumption. Hence, for all primes v which are ramified in LM_λ, we have $D_v = I_v$ as desired.

Second part: the case where $\tilde{G}$ is a non-split extension

The proof will be carried out in three stages:

(i) Existence of an extension $\tilde{L}$ giving a solution to the embedding problem.

(ii) Modifying $\tilde{L}$ so that it is ramified at the same places as L.

(iii) Modifying $\tilde{L}$ further so that it has property (S_N), with at most one additional ramified prime.

(i) Solvability of the embedding problem

The field extension L determines a surjective homomorphism $\phi : G_{\mathbf{Q}} \to G$. The problem is to lift ϕ to a homomorphism $\tilde{\phi} : G_{\mathbf{Q}} \to \tilde{G}$. (Such a $\tilde{\phi}$ is automatically surjective because of our assumption that $\tilde{G}$ does not split.) Let $\xi \in H^2(G, C_l)$ be the class of the extension $\tilde{G}$, and let

$$\phi^* : H^2(G, C_l) \longrightarrow H^2(G_{\mathbf{Q}}, C_l)$$

be the homomorphism defined by ϕ. The existence of the lifting $\tilde{\phi}$ is equivalent to the vanishing of $\phi^*(\xi)$ in $H^2(G_{\mathbf{Q}}, C_l)$. As usual in Galois cohomology, we write $H^2(G_{\mathbf{Q}}, -)$ as $H^2(\mathbf{Q}, -)$, and similarly for other fields. The following well-known lemma reduces the statement $\phi^* \xi = 0$ to a purely local question:

Lemma 2.1.5 *The restriction map*

$$H^2(\mathbf{Q}, C_l) \longrightarrow \prod_p H^2(\mathbf{Q}_p, C_l)$$

is injective.

(A similar result holds for any number field.)

Sketch of Proof. Let $K = \mathbf{Q}(\mu_l)$. Since $[K : \mathbf{Q}]$ is prime to l, the map $H^2(\mathbf{Q}, C_l) \longrightarrow H^2(K, \mathbf{C}_l)$ is injective. Hence, it is enough to prove the lemma with $\mathbf{Q}$ replaced by K. In that case, $H^2(K, C_l)$ is isomorphic to $\mathrm{Br}_l(K)$, the l-torsion of the Brauer group of K. The lemma then follows from the Brauer-Hasse-Noether theorem: an element of $\mathrm{Br}(K)$ which is 0 locally is 0. (Note that, since $l \neq 2$, the archimedean places can be ignored.)

By the above lemma, it suffices to show that $\phi^* \xi = 0$ locally at all primes. In other words, we must lift the map $\phi_p : G_{\mathbf{Q}_p} \longrightarrow D_p \subset G$ to $\tilde{\phi}_p : G_{\mathbf{Q}_p} \longrightarrow \tilde{G}$. There are two cases:

1. p is unramified in L, i.e., ϕ_p is trivial on the inertia group I_p of $G_{\mathbf{Q}_p}$. Then ϕ_p factors through the quotient $G_{\mathbf{Q}_p}/I_p = \hat{\mathbf{Z}}$. But one can always lift a map $\hat{\mathbf{Z}} \longrightarrow G$ to a map $\hat{\mathbf{Z}} \longrightarrow \tilde{G}$: just lift the generator of $\hat{\mathbf{Z}}$.

2. p is ramified in L. By construction, $p \equiv 1 \pmod{l^N}$, hence $p \neq l$ and $L_v/\mathbf{Q}_p$ is tamely ramified (as in 2.1.2, v denotes a place of L above

p); since its Galois group D_v is equal to its inertia group I_v, it is cyclic. The homomorphism $G_{\mathbf{Q}_p} \longrightarrow D_v \subset G$ factors through the map $G_{\mathbf{Q}_p} \longrightarrow \mathrm{Gal}(E/\mathbf{Q}_p)$, where E is the maximal abelian tame extension of $\mathbf{Q}_p$ with exponent dividing l^N. The extension E can be described explicitly: it is composed of the unique unramified extension of $\mathbf{Q}_p$ of degree l^N, (obtained by taking the fraction field of the ring of Witt vectors over $\mathbf{F}_{p^{lN}}$) and the totally ramified extension $\mathbf{Q}_p(\sqrt[l^N]{p})$ (which is a Kummer extension since $p \equiv 1 \pmod{l^N}$). It follows that $\mathrm{Gal}(E/\mathbf{Q}_p)$ is an abelian group of type (l^N, l^N); it is projective in the category of abelian groups of exponent dividing l^N. The inverse image of D_v in $\tilde{G}$ belongs to that category (a central extension of a cyclic group is abelian). This shows that the local lifting is possible.

(ii) Modifying the extension $\tilde{L}$ so that it becomes unramified outside the set $\mathrm{ram}(L/\mathbf{Q})$ of primes ramified in $L/\mathbf{Q}$

Lemma 2.1.6 *For every prime p, let ϵ_p be a continuous homomorphism from $\mathrm{Gal}(\bar{\mathbf{Q}}_p/\mathbf{Q}_p)$ to a finite abelian group C. Suppose that almost all ϵ_p are unramified. Then there is a unique $\epsilon : \mathrm{Gal}(\bar{\mathbf{Q}}/\mathbf{Q}) \longrightarrow C$, such that for all p, the maps ϵ and ϵ_p agree on the inertia groups I_p.*

(The decomposition and inertia groups D_p, I_p are only defined up to conjugacy inside $G_{\mathbf{Q}}$. We shall implicitly assume throughout that a fixed place v has been chosen above each p, so that D_p and I_p are well-defined subgroups of $G_{\mathbf{Q}}$.)

Proof of lemma: By local class field theory, the ϵ_p can be canonically identified with maps $\mathbf{Q}_p^* \longrightarrow C$. The restrictions of ϵ_p to $\mathbf{Z}_p^*$ are trivial on a closed subgroup $1 + p^{n_p}\mathbf{Z}_p$, where n_p is the *conductor* of ϵ_p. Since almost all n_p are zero, there is a homomorphism $\epsilon : (\mathbf{Z}/M\mathbf{Z})^* \longrightarrow C$, with $M = \prod p^{n_p}$, and $\epsilon(k) = \prod \epsilon_p(k^{-1})$. If we view ϵ as a Galois character, class field theory shows that it has the required properties. (Equivalently, one may use the direct product decomposition of the idèle group $I_{\mathbf{Q}}$ of $\mathbf{Q}$, as:

$$I_{\mathbf{Q}} = \left(\prod_p \mathbf{Z}_p^* \times \mathbf{R}_+^* \right) \times \mathbf{Q}^*.)$$

Proposition 2.1.7 *Let $1 \to C \to \tilde{\Phi} \to \Phi \to 1$ be a central extension of a group Φ, and ϕ be a continuous homomorphism from $G_{\mathbf{Q}}$ to Φ which has a lifting $\psi : G_{\mathbf{Q}} \longrightarrow \tilde{\Phi}$. Let $\tilde{\phi}_p : G_{\mathbf{Q}_p} \longrightarrow \tilde{\Phi}$ be liftings of $\phi_p = \phi|_{D_p}$, such that the $\tilde{\phi}_p$ are unramified for almost all p. Then there is a lifting $\tilde{\phi} : G_{\mathbf{Q}} \longrightarrow \tilde{\Phi}$ such that, for every p, $\tilde{\phi}$ is equal to $\tilde{\phi}_p$ on the inertia group at p. Such a lifting is unique.*

This proposition is also useful for relating Galois representations in $\mathbf{GL}_n$ and $\mathbf{PGL}_n$ (Tate, see [Se7, §6]).)

Proof of prop. 2.1.7: For every p, there is a unique homomorphism

$$\epsilon_p : G_{\mathbf{Q}_p} \longrightarrow C$$

such that

$$\psi(s) = \epsilon_p(s)\tilde{\phi}_p(s)$$

for all $s \in G_{\mathbf{Q}_p}$. By the previous lemma, there exists a unique $\epsilon : G_{\mathbf{Q}} \longrightarrow C$ which agrees with ϵ_p on I_p. The homomorphism $\phi = \psi\epsilon^{-1}$ has the required property. This proves the existence assertion. The uniqueness is proved similarly.

Corollary 2.1.8 *Assuming the hypotheses of prop. 2.1.7, a lifting of ϕ can be chosen unramified at every prime where ϕ is unramified.*

Proof: Choose local liftings $\tilde{\phi}_p$ of ϕ which are unramified where ϕ is; this is possible since there is no obstruction to lifting a homomorphism defined on $\hat{\mathbf{Z}}$. Then, apply prop. 2.1.7.

The corollary completes the proof of part (ii): $\tilde{L}$ can be modified so that it is ramified at the same places as L.

(iii) Modifying $\tilde{L}$ so that it satisfies property (S_N)

We have obtained an extension $\tilde{L}$ which is ramified at the same places as L and which solves the extension problem for $\tilde{G}$. Let p be in $\mathrm{ram}(L/\mathbf{Q}) = \mathrm{ram}(\tilde{L}/\mathbf{Q})$. Denote by D_p, I_p (resp $\tilde{D}_p$, $\tilde{I}_p$) the decomposition and inertia groups for L (resp $\tilde{L}$) at p. We have $I_p = D_p \subset G$; this is a cyclic group of order l^α, say. Let I'_p be the inverse image of I_p in $\tilde{G}$. We have $\tilde{I}_p \subset \tilde{D}_p \subset I'_p$. If I'_p is a non-split extension of I_p (i.e., is cyclic of order $l^{\alpha+1}$) we even have $\tilde{I}_p = \tilde{D}_p = I'_p$, and the Scholz condition is satisfied at p. Let S be the set of $p \in \mathrm{ram}(L/\mathbf{Q})$ for which I'_p is a split extension of I_p; since $\tilde{I}_p$ is cyclic, we have $I'_p = \tilde{I}_p \times C_l$. The Frobenius element $\mathrm{Frob}_p \in \tilde{D}_p/\tilde{I}_p \subset I'_p/\tilde{I}_p$ may be identified with an element c_p of C_l; the Scholz condition is satisfied at p if and only if $c_p = 1$. If all c_p's are equal to 1, $\tilde{L}$ satisfies (S_N). If not, we need to correct $\tilde{\phi} : G_{\mathbf{Q}} \longrightarrow \tilde{G}$ by a Galois character $\chi : (\mathbf{Z}/q\mathbf{Z})^* \longrightarrow C_l$ which satisfies the following properties:

1. $q \equiv 1 \pmod{l^N}$.
2. For every p in S, $\chi(p) = c_p$.
3. The prime q splits completely in $L/\mathbf{Q}$.

Conditions 1, 2, and 3 impose conditions on the behaviour of q in the fields $\mathbf{Q}(\sqrt[l^N]{1})$, $\mathbf{Q}(\sqrt[l]{1}, \sqrt[l]{p}, p \in S)$, and L respectively. Write $\mathbf{Q}(\sqrt[l^N]{1}) = \mathbf{Q}(\sqrt[l]{1}) \cdot F$, where F is cyclic of order l^{N-1} and totally ramified at l.

Lemma 2.1.9 *The fields L, F, and $\mathbf{Q}(\sqrt[l]{1}, \sqrt[l]{p}, p \in S)$ are linearly disjoint over $\mathbf{Q}$.*

Proof: Since L and F have distinct ramification, L and F are linearly disjoint: $L \cdot F$ has Galois group $G \times C_{l^{N-1}}$. The extension $\mathbf{Q}(\sqrt[l]{1}, \sqrt[l]{p}, p \in S)$ has Galois group $V = C_l \times C_l \ldots \times C_l$ ($|S|$ times) over $\mathbf{Q}(\sqrt[l]{1})$. The action of $\mathrm{Gal}(\mathbf{Q}(\sqrt[l]{1})/\mathbf{Q}) = \mathbf{F}_l^*$ on V by conjugation is the natural action of multiplication by scalars. The Galois group of $\mathbf{Q}(\sqrt[l]{1}, \sqrt[l]{p}, p \in S)$ over $\mathbf{Q}$ is a semi-direct product of $\mathbf{F}_l^*$ with V. Since $l \neq 2$, this group has no quotient of order l: there is no Galois subfield of $\mathbf{Q}(\sqrt[l]{1}, \sqrt[l]{p}, p \in S)$ of degree l over $\mathbf{Q}$. This implies that $L \cdot F$ and $\mathbf{Q}(\sqrt[l]{1}, \sqrt[l]{p}, p \in S)$ are linearly disjoint. QED.

If $S = \{p_1, \ldots, p_k\}$, define integers ν_i, $2 \leq i \leq k$, by $c_{p_i} = c_{p_1}^{\nu_i}$. (This is possible if $c_1 \neq 1$, which we may assume.) In order to satisfy conditions $1, 2$ and 3, the prime q must have the following behaviour in the extension $L \cdot F \cdot \mathbf{Q}(\sqrt[l]{1}, \sqrt[l]{p}, p \in S)$:

$$\left\{ \begin{array}{ll} \mathrm{Frob}_q = 1 & \text{in } L \cdot F \text{ and } \mathbf{Q}(\sqrt[l^N]{1}); \\ \mathrm{Frob}_q \neq 1 & \text{in } \mathbf{Q}(\sqrt[l]{1}, \sqrt[l]{p_1}); \\ \mathrm{Frob}_q = 1 & \text{in } \mathbf{Q}(\sqrt[l]{1}, \sqrt[l]{p_1/p_i^{\nu_i}}), \ i = 2, \ldots, k. \end{array} \right.$$

By the Chebotarev density theorem and lemma 2.1.9, such a q exists. One can then define the character χ so that $\chi(p_i) = c_{p_i}$. This completes part (ii): the homomorphism $\tilde{\phi}\chi^{-1}$ defines a new $\tilde{G}$-extension $\tilde{L}$ with property (S_N), and with one additional ramified prime, namely q.

The proof allows us to generalize the theorem somewhat. Let us make the following definition:

Definition 2.1.10 *If G is a profinite group, the following are equivalent:*
1. The topology of G is metrizable.
2. G can be written as a denumerable projective limit

$$G = \varprojlim(\cdots \rightarrow G_n \rightarrow G_{n-1} \rightarrow \cdots),$$

where the G_n's are finite (and the connecting homomorphisms are surjective).
3. The set of open subgroups of G is denumerable.

A group G which satisfies these equivalent properties is said to be *separable*.

If $G = \mathrm{Gal}(L/K)$, these properties are equivalent to $[L : K] \leq \aleph_0$; if G is a pro-l-group, they are equivalent to $\dim H^1(G, \mathbf{Z}/l\mathbf{Z}) \leq \aleph_0$.

The proof that the three properties in the definition are equivalent is elementary.

Theorem 2.1.11 *If G is a separable pro-l-group of finite exponent, then there is a Galois extension of $\mathbf{Q}$ with Galois group G.*

Proof: If l^N is the exponent of G, write G as $\mathrm{proj.lim}(G_n)$ where each G_n is a finite l-group, the connecting homomorphism being surjective, with kernel of order l. By th. 2.1.3, one can construct inductively an increasing family of Galois extensions $L_n/\mathbf{Q}$ with Galois group G_n which have the (S_N) property; the union of the L_n's has Galois group G.

Remark: The finiteness condition on the exponent cannot be dropped: for example, $\mathbf{Z}_l \times \mathbf{Z}_l$ is not a Galois group over $\mathbf{Q}$.

A more general result has been proved by Neukirch [Ne] for pro-solvable groups of odd order and finite exponent.

2.2   The Frattini subgroup of a finite group

Let G be a finite group.

Definition 2.2.1 *The* Frattini subgroup Φ *of G is the intersection of the maximal subgroups of G.*

The Frattini subgroup is normal. If $G \neq 1$, then $\Phi \neq G$. If $G_1 \subset G$ satisfies $\Phi \cdot G_1 = G$, then $G_1 = G$. (Otherwise, choose a maximal subgroup M such that $G_1 \subset M \subset G$. Since $\Phi \subset M$, it follows that $\Phi G_1 \subset M$, which is a contradiction.) In other words, a subset of G generates G if and only if it generates G/Φ: elements of Φ are sometimes referred to as "non-generators".

Examples:
1. If G is a simple group, then $\Phi = 1$.
2. If G is a p-group, the maximal subgroups are the kernels of the surjective homomorphisms $G \longrightarrow C_p$. Hence Φ is generated by (G, G) and G^p, where (G, G) denotes the commutator subgroup of G; more precisely, we have

$$\Phi = (G, G) \cdot G^p.$$

The group G/Φ is the maximal abelian quotient of G of type $(p, p, \ldots, p)$.

Proposition 2.2.2 ([Hu], p. 168) *Let G be a finite group, Φ its Frattini subgroup, N a normal subgroup of G with $\Phi \subset N \subset G$. Assume N/Φ is nilpotent. Then N is nilpotent.*

Corollary 2.2.3 *The group Φ is nilpotent.*

This follows by applying prop. 2.2.2 to $N = \Phi$.

Let us prove prop. 2.2.2. Recall that a finite group is nilpotent if and only if it has only one Sylow p-subgroup for every p. Choose a Sylow p-subgroup P of N, and let $Q = \Phi P$. The image of Q by the quotient map $N \longrightarrow N/\Phi$ is a Sylow p-subgroup of N/Φ which is unique by assumption. Hence this image is a characteristic subgroup of N/Φ; in particular it is preserved by inner conjugation by elements of G, i.e., Q is normal in G. Let

$$N_G(P) = \{g | g \in G, gPg^{-1} = P\}$$

be the normalizer of P in G. If $g \in G$, then gPg^{-1} is a Sylow p-subgroup of Q. Applying the Sylow theorems in Q, there is a $q \in Q$ such that

$$qgPg^{-1}q^{-1} = P.$$

Hence $qg \in N_G(P)$. It follows that $G = QN_G(P) = \Phi N_G(P)$. Therefore $G = N_G(P)$, and P is normal in G, hence in N; this implies that P is the only Sylow p-subgroup of N.

Application to solvable groups

Proposition 2.2.4 *Let G be a finite solvable group $\neq 1$. Then G is isomorphic to a quotient of a group H which is a semi-direct product $U \cdot S$, where U is a nilpotent normal subgroup of H, and S is solvable with $|S| < |G|$.*

Proof: Let Φ be the Frattini subgroup of G; since G/Φ is solvable and $\neq 1$, it contains a non-trivial abelian normal subgroup, e.g., the last non-trivial term of the descending derived series of G/Φ. Denote by U its inverse image in G. Since $\Phi \subset U \subset G$, with U/Φ abelian, U is nilpotent by prop. 2.2.2. Choose a maximal subgroup S of G which does not contain U: this is possible since $U \neq \Phi$. Since $U \cdot S \neq S$ and S is maximal, $G = U \cdot S$. Hence, writing $H = U \cdot S$ (with S acting by conjugation on the normal subgroup U), there is a surjective map $H \longrightarrow G$.

The relevance of prop. 2.2.4 to Galois theory lies in the following result which asserts that the embedding problem for split extensions with nilpotent kernel has always a solution.

Claim 2.2.5 *([Sha2], [Is]) Let L/K be an extension of number fields with Galois group S, let U be a nilpotent group with S-action, and let G be the semi-direct product $U \cdot S$. Then the embedding problem for L/K and for*

$$1 \to U \to G \to S \to 1$$

has a solution.

Theorem 2.2.6 *Claim 2.2.5 implies the existence of Galois extensions of* **Q** *with given solvable Galois group.*

Proof: Let G be a solvable group. We proceed by induction on the order of G. We may asume $G \neq 1$. By prop. 2.2.4, write G as a quotient of $U \cdot S$ with U nilpotent and S solvable, $|S| < |G|$. The induction hypothesis gives a Galois extension $L/\mathbf{Q}$ with Galois group S. By the claim above, $U \cdot S$ can be realized as a Galois group; hence, so can its quotient G.

Let us give a proof of claim 2.2.5 in the elementary case where U is abelian of exponent n. Observe that:

1. If claim 2.2.5 is true for an extension L' of L, it is true for L: for, if $S' = \mathrm{Gal}(L/K)$, there is a natural quotient map $US' \to US$. Hence we may assume $\mu_n \subset L$, where μ_n denotes the nth roots of unity.

2. We may also assume

$$U \simeq \text{ direct sum of copies of } \mathbf{Z}/n\mathbf{Z}[S],$$

because any abelian group of exponent n on which S acts is a quotient of such an S-module.

Suppose that h is the number of copies in the decomposition of U as a direct sum of S-modules $\mathbf{Z}/n\mathbf{Z}[S]$. Choose places $v_1, \ldots, v_h$ of K which split completely in L, $w_1, \ldots, w_h$ places of L which extend them; any place of L extending one of the v_i can be written uniquely as sw_i, for some $s \in S$. Choose elements $\phi_j \in L^*$ such that

$$(sw_i)(\phi_j) = \begin{cases} 1 & \text{if } s = 1 \text{ and } i = j. \\ 0 & \text{otherwise.} \end{cases}$$

Let M be the field generated over L by the $\sqrt[n]{s\phi_j}$, for $s \in S$ and $j = 1, \ldots, h$. This is a Galois extension of K, with $\mathrm{Gal}(M/L) \simeq U$. Its Galois group over K is an extension of S by U; since U is a free $\mathbf{Z}/n\mathbf{Z}[S]$-module, it is known that such an extension splits (see, e.g., [Se2, ch. IX]). Hence $\mathrm{Gal}(M/K)$ is isomorphic to the semi-direct product of S by U.

Chapter 3

Hilbert's irreducibility theorem

3.1 The Hilbert property

Fix a ground field K with Char $K = 0$, and let V be an irreducible algebraic variety over K. (In what follows, algebraic varieties will be tacitly assumed to be integral and quasi-projective.) Denote by $V(K)$ the set of K-rational points of V.

A subset A of $V(K)$ is said to be of *type* (C_1) if there is a closed subset $W \subset V$, $W \neq V$, with $A \subset W(K)$, i.e., if A is not Zariski-dense in V.

A subset A of $V(K)$ is said to be of *type* (C_2) if there is an irreducible variety V', with $\dim V = \dim V'$, and a generically surjective morphism $\pi : V' \longrightarrow V$ of degree ≥ 2, with $A \subset \pi(V'(K))$.

Definition 3.1.1 *A subset A of $V(K)$ is called* thin *("mince" in French) if it is contained in a finite union of sets of type (C_1) or (C_2).*

Alternately, a set A is thin if there is a morphism

$$\pi : W \to V \text{ with } \dim W \leq \dim V$$

having no rational cross-section, and such that $A \subset \pi(W(K))$.

Example: If $V = \mathbf{P}_1$, $V(K) = K \cup \{\infty\}$. The set of squares (resp. cubes,...) in K is thin.

Definition 3.1.2 (cf. [CTS1], p. 189) *A variety V over K satisfies the Hilbert property if $V(K)$ is not thin.*

This is a birational property of V.

Definition 3.1.3 *A field K is* Hilbertian *if there exists an irreducible variety V over K, with $\dim V \geq 1$, which has the Hilbert property.*

It is easy to show (cf. exerc. 1) that if K is Hilbertian, then the projective line $\mathbf{P}_1$ over K has the Hilbert property (hence, our definition is equivalent to the standard one, see e.g. [L]).

The fields $\mathbf{R}, \mathbf{Q}_p$ are not Hilbertian. A number field is Hilbertian (see §3.4).

Remark on irreducible varieties: The variety V is said to be *absolutely irreducible* if the algebraic closure K' of K in the field $K(V)$ of rational functions on V is equal to K. Equivalently, V must remain irreducible upon extension of scalars to the algebraic closure $\bar{K}$ of K. If V is not absolutely irreducible, then $V(K) \subset W(K)$, where W is a subvariety of V, $W \neq V$. Indeed, if V is a normal variety, then $V(K) = \emptyset$. For, the residue field of the local ring at $P \in V$ contains K', and hence no point of V is K-rational. The general case follows from this by normalization. In particular, an irreducible variety which has the Hilbert property is absolutely irreducible.

Therefore, in our definition of C_2-type subsets, we could have asked that V' be absolutely irreducible.

Remark: Let $\pi : V' \longrightarrow V$ be a finite morphism (we also say that π is a "covering", even though it can be ramified). Assume that V and V' are absolutely irreducible, and let $K(V')/K(V)$ be the corresponding field extension. Let $K(V')^{\text{gal}}$ be the Galois closure of $K(V')$ over $K(V)$, and let W be the normalisation of V' in $K(V')^{\text{gal}}$. The variety W with its projection $W \longrightarrow V$ may be called the *Galois closure* of $V' \longrightarrow V$. Note that K is not always algebraically closed in $K(V')^{\text{gal}}$, i.e., W need not be absolutely irreducible. For example, take $V = V' = \mathbf{P}_1$, $\pi(x) = x^3$; the Galois closure of V' is $\mathbf{P}_{1/K(\mu_3)}$, and hence is not absolutely irreducible over K if K does not contain μ_3.

Exercises:
1. Let V be an affine irreducible variety over K, with $\dim V \geq 1$. Let $W_1, \ldots, W_r$ be absolutely irreducible coverings of the projective line $\mathbf{P}_1$. Show that there exists a morphism $f : V \longrightarrow \mathbf{P}_1$ such that the pullback coverings f^*W_i of V are absolutely irreducible. Use this to show that if V has the Hilbert property, then so has $\mathbf{P}_1$.
2. Let V and T be absolutely irreducible varieties, and $A \subset V(K)$ a thin subset. Show that $A \times T(K)$ is thin in $V \times T$. More generally, let $f : W \longrightarrow V$ be a generically surjective morphism whose generic fiber is absolutely irreducible (i.e. the function field extension $K(W)/K(V)$ is regular). If B is a subset of $W(K)$ such that $f(B)$ is thin in $V(K)$, show that B is thin in $W(K)$.

3. Let K be a Hilbertian field, and let A be the set of elements of K which are sums of two squares. Show that A is not thin. (Use exerc. 2.)

4. Let K be a number field and V an abelian variety over K with $\dim V \geq 1$. Show that $V(K)$ is thin (i.e., V does not have the Hilbert property). Hint: use the Mordell-Weil theorem.

Problem: If V and V' are irreducible varieties with the Hilbert property, is it true that $V \times V'$ has the same property?

3.2 Properties of thin sets

3.2.1 Extension of scalars

Let L/K be a finite extension, V an absolutely irreducible variety over K. Extension of scalars to L yields a variety over L, denoted $V_{/L}$.

Proposition 3.2.1 *If $A \subset V(L)$ is thin with respect to L, then $A \cap V(K)$ is thin with respect to K.*

The proof uses the restriction of scalars functor $R_{L/K} : (\mathrm{Var}_L) \longrightarrow (\mathrm{Var}_K)$ from L-varieties to K-varieties, cf. [We], [Oe]. Here are two equivalent definitions of $R_{L/K}$:

1. It is the right adjoint to the extension of scalars $(\mathrm{Var}_K) \longrightarrow (\mathrm{Var}_L)$, i.e., for every K-variety T and L-variety W, one has:

$$\mathrm{Mor}_K(T, R_{L/K}W) = \mathrm{Mor}_L(T_{/L}, W);$$

In particular, taking T to be a point which is a rational over K, the above formula yields

$$(R_{L/K}W)(K) = W(L).$$

2. Let Σ_L be the set of embeddings of L in some fixed algebraic closure $\bar{K}$; for each $\sigma \in \Sigma_L$, let W^σ be the variety deduced from the given L-variety W by extension of scalars via σ. Then the product $X = \prod_\sigma W^\sigma$ is a $\bar{K}$-variety. Moreover, one has natural isomorphisms from X to X^s for every $s \in G_K$. By Weil's descent theory, these isomorphisms give rise to a K-variety from which X comes by extension of scalars; this variety is $R_{L/K}W$.

If A is of type (C_1), then $A \cap V(K)$ is clearly of type (C_1). Hence we may assume that $A \subset \pi(W(L))$, that W is absolutely irreducible over L with $\dim W = \dim V$, and π is a covering $W \longrightarrow V$ with $\deg \pi > 1$. By restricting suitably V, we may assume that π is finite étale. The functor $R_{L/K}$ then gives an étale covering $R_{L/K}W \longrightarrow R_{L/K}V_{/L}$. Using the diagonal embedding $\Delta : V \longrightarrow R_{L/K}V_L$, we obtain an étale covering $\pi' : V' \longrightarrow V$, and a Cartesian diagram:

$$
\begin{array}{ccc}
V' & \longrightarrow & R_{L/K}W \\
\downarrow{\scriptstyle \pi'} & & \downarrow \\
V & \xrightarrow{\Delta} & R_{L/K}V_{/L}.
\end{array}
$$

The set $A \cap V(K)$ is contained in $\pi'(V'(K))$, and it is easy to check that all the components of V' have degree over V at least equal to $\deg \pi$. Hence $\pi'(V'(K))$ is thin, and the same is true for $A \cap V(K)$.

Corollary 3.2.2 *If K is Hilbertian, then so is L.*

Suppose L were not. Then $A = \mathbf{P}_1(L)$ is thin in $\mathbf{P}_1$ with respect to L. This implies that $A \cap \mathbf{P}_1(K) = \mathbf{P}_1(K)$ is thin in $\mathbf{P}_1$ with respect to K; contradiction.

Remark: The converse to cor. 3.2.2 is not true; see e.g. [Ku].

3.2.2   Intersections with linear subvarieties

Let V be the projective space $\mathbf{P}_n$ of dimension n, and let $A \subset V(K)$ be a thin set. We denote by $\mathbf{Grass}_n^d$ the Grassmann variety of d-linear subspaces of $\mathbf{P}_n$, where $1 \leq d \leq n$.

Proposition 3.2.3 *There is a non-empty Zariski-open subset $U \subset \mathbf{Grass}_n^d$ such that if W belongs to $U(K)$, then $A \cap W$ is thin in W.*

It is enough to prove this when A is either of type (C_1) or of type (C_2). The first case is easy. In the second, there is a map $\pi : V' \longrightarrow \mathbf{P}_n$, with V' absolutely irreducible, $\deg \pi \geq 2$, and $A \subset \pi(V'(K))$. By Bertini's theorem (see e.g., [Jou, ch. I, §6], [Ha, p. 179], [De2], [Z]), there exists a non-empty open set U in $\mathbf{Grass}_n^d$ such that $\pi^{-1}(W)$ is absolutely irreducible for all $W \in U$. Hence, if $W \in U(K)$, then $W \cap A$ is of type (C_2), and hence is thin.

An interesting case occurs when $d = 1$. Let $\pi : V' \longrightarrow \mathbf{P}_n$ be a generically surjective map of degree > 1, and Φ the hypersurface of ramification of π. Consider the set U of lines which intersect Φ transversally at smooth points. Then for $\mathcal{L} \in U$, the covering $\pi^{-1}(\mathcal{L}) \longrightarrow \mathcal{L}$ is irreducible: one proves this over $\mathbf{C}$ by deforming the line into a generic one, and the general case follows.

Example: Consider the "double plane" V'_F with equation $t^2 = F(x, y)$, where F is the equation for a smooth quartic curve Φ in $\mathbf{P}_2$. The natural projection of V_F' onto $\mathbf{P}_2$ is quadratic and ramified along the curve Φ. A line in $\mathbf{P}_2$ which intersects Φ transversally in 4 points lifts to an irreducible curve of genus 1 in V_F'; a line which is tangent at one point and at no other lifts to an irreducible curve of genus zero; finally, if the line is one of the 28 bitangents to Φ, then its inverse image is two curves of genus zero. In that case, one can take for U the complement of 28 points in $\mathbf{P}_2 = \mathbf{Grass}_2^1$.

Corollary 3.2.4 *If $\mathbf{P}_n$ has the Hilbert property over K for some $n \geq 1$, then all projective spaces $\mathbf{P}_m$ over K have the Hilbert property.*

Proof: $\mathbf{P}_1$ has the Hilbert property over K: if not, $\mathbf{P}_1(K) = A$ is thin, and $A \times \mathbf{P}_1(K) \times \ldots \times \mathbf{P}_1(K)$ is thin in $\mathbf{P}_1 \times \ldots \times \mathbf{P}_1$. This cannot be the case, since $\mathbf{P}_n$ has the Hilbert property (and hence also $\mathbf{P}_1 \times \cdots \times \mathbf{P}_1$ which is birationally isomorphic to $\mathbf{P}_n$). This implies the same for $\mathbf{P}_m$, with $m \geq 1$. For, if $A = \mathbf{P}_m(K)$ is thin in $\mathbf{P}_m$, then by prop. 3.2.3, there is a line $\mathcal{L}$ such that $\mathcal{L} \cap A = \mathcal{L}(K)$ is thin in $\mathcal{L} = \mathbf{P}_1$. But this contradicts the fact that $\mathbf{P}_1(K)$ has the Hilbert property.

3.3 Irreducibility theorem and thin sets

Let $\pi : W \longrightarrow V$ be a Galois covering with Galois group G, where V, W denote K-irreducible varieties, $V = W/G$, and G acts faithfully on W. Let us say that $P \in V(K)$ has property $\mathrm{Irr}(P)$ if P is "inert", i.e., the inverse image of P (in the scheme sense) is one point, i.e., the affine ring of the fiber is a field K_P (or, equivalently, G_K acts freely and transitively on the $\bar{K}$-points of W above P). In this case, π is étale above P and the field K_P is a Galois extension of K with Galois group G.

Proposition 3.3.1 *There is a thin set $A \subset V(K)$ such that for all $P \notin A$, the irreducibility property $\mathrm{Irr}(P)$ is satisfied.*

Proof: By removing the ramification locus, we may assume that $W \longrightarrow V$ is étale, i.e., G acts freely on each fiber. Let Σ be the set of proper subgroups H of G. We denote by W/H the quotient of W by H, and by π_H the natural projection onto V. Let

$$A = \bigcup_{H \in \Sigma} \pi_H(W/H)(K).$$

The set A is thin, since the degrees of the π_H are equal to $[G : H] > 1$. If $P \notin A$, then $\mathrm{Irr}(P)$ is satisfied: for, lift P to $\bar{P}$ in $W(\bar{K})$, and let H be the subgroup of G consisting of elements $g \in G$ such that $g\bar{P} = \gamma\bar{P}$ for some $\gamma \in G_K = \mathrm{Gal}(\bar{K}/K)$. Then $H = G$. Otherwise, H would belong to Σ, and since the image of $\bar{P}$ in W/H is rational over K, the point P would be in A.

Corollary 3.3.2 *If V has the Hilbert property over K, the existence of a G-covering $W \longrightarrow V$ as above implies that there is a Galois extension of K with Galois group G.*

Assume furthermore that W, and hence V, are absolutely irreducible, and that V has the Hilbert property. Let us say that an extension L/K is of *type W* if it comes from lifting a K-rational point on V to W.

Proposition 3.3.3 *Under these assumptions, for every finite extension L of K, there is a Galois extension E/K of type W with Galois group G which is linearly disjoint from L.*

Corollary 3.3.4 *There exist infinitely many linearly disjoint extensions with Galois group G, of type W.*

Proof of prop. 3.3.3: Extend scalars to L; applying prop. 3.3.1 there is a thin set $A_L \subset V_{/L}(L)$ such that for all $P \notin A$, the property $\mathrm{Irr}(P)$ is satisfied over L. Set $A = A_L \cap V(K)$. This is a thin set by prop. 3.2.1. Choose $P \in V(K)$ with $P \notin A$. Then $\mathrm{Irr}(P)$ is true over L, and hence *a fortiori* over K. This P gives the desired extension E.

Exercise: Show that the set of rational points P satisfying property $\mathrm{Irr}(P)$ and giving a fixed Galois extension of K is thin (if $G \neq 1$).

Polynomial interpretation

Let V be as above, let $K(V)$ be its function field, and let

$$f(X) = X^n + a_1 X^{n-1} + \cdots + a_n, \quad a_i \in K(V)$$

be an irreducible polynomial over $K(V)$. Let $G \subset S_n$ be the Galois group of f viewed as a group of permutations on the roots of f. (This group can be identified with a subgroup of S_n, up to conjugacy in S_n, i.e., one needs to fix a labelling of the roots.) If $t \in V(K)$ and t is not a pole of any of the a_i, then $a_i(t) \in K$, and one can define the specialization of f at t:

$$f_t(X) = X^n + a_1(t)X^{n-1} + \cdots + a_n(t).$$

Proposition 3.3.5 *There exists a thin set $A \subset V(K)$ such that, if $t \notin A$, then:*

    1. t is not a pole of any of the a_i,
    2. $f_t(X)$ is irreducible over K,
    3. the Galois group of f_t is G.

By replacing V by a dense open subset, we may assume that the a_i have no poles, that V is smooth and that the discriminant Δ of f is invertible. The subvariety V_f of $V \times \mathbf{A}^1$ defined by

$$(t, x) \in V_f \iff f_t(x) = 0$$

is an étale covering of degree n. Its Galois closure $W = V_f^{\mathrm{gal}}$ has Galois group G. The proposition follows by applying prop. 3.3.1 to W.

Examples:

• $G = S_3$. Let $f(X) = X^3 + a_1 X^2 + a_2 X + a_3$ be irreducible, with Galois group S_3 over $K(V)$. The specialization at t has Galois group S_3 if the following properties are satisfied:

    1. $t \in V(K)$ is not a pole for any of the a_i.

    2. $\Delta(t) \neq 0$ (where $\Delta = a_1^2 a_2^2 + 18 a_1 a_2 a_3 - 4 a_1^3 a_3 - 4 a_2^3 - 27 a_3^2$ is the discriminant of f).

    3. $X^3 + a_1(t) X^2 + a_2(t) X + a_3(t)$ has no root in K.

    4. $\Delta(t)$ is not a square in K.

Conditions 3 and 4 guarantee that G is not contained in either of the maximal subgroups of S_3, namely S_2 and A_3. It is clear that the set A of $t \in V(K)$ which fail to satisfy these conditions is thin.

• $G = S_4$. The maximal subgroups of G are S_3, A_4, and D_4, the dihedral group of order 8. The case of $G \subset S_3$ or A_4 can be disposed of by imposing the same conditions as in the case $G = S_3$; to handle the case of D_4, one requires the cubic resolvent

$$(X - (x_1 x_2 + x_3 x_4))(X - (x_1 x_3 + x_2 x_4))(X - (x_1 x_4 + x_2 x_3))$$
$$= X^3 - a_2 X^2 + (a_1 a_3 - 4 a_4) X + (4 a_2 a_4 - a_1^2 a_4 - a_3^2)$$

to have no root in K, where $x_1, \ldots, x_4$ are the roots of the polynomial

$$f = X^4 + a_1 X^3 + \ldots + a_4.$$

• $G = S_5$. The maximal subgroups of G are S_4, $S_2 \times S_3$, and A_5, which give conditions similar to the above, and the Frobenius group F_{20} of order 20, which is a semi-direct product $C_5 C_4$ and can be viewed as the group of affine linear transformations of the form $x \mapsto ax + b$ on $\mathbf{Z}/5\mathbf{Z}$. If the Galois group of the specialized polynomial is contained in F_{20}, then the sextic resolvent, which is the minimal polynomial over $K(V)$ for

$$\frac{(x_1 x_2 + x_2 x_3 + x_3 x_4 + x_4 x_5 + x_5 x_1) - (x_1 x_3 + x_2 x_4 + x_3 x_5 + x_4 x_1 + x_5 x_2)}{\prod_{i<j}(x_i - x_j)},$$

has a root in K when specialized.

3.4   Hilbert's irreducibility theorem

Theorem 3.4.1 (Hilbert [Hi]) *If K is a number field, then for every n, the affine space $\mathbf{A}^n$ (or equivalently the projective space $\mathbf{P}_n$) has the Hilbert property over K. (In other words, K is Hilbertian.)*

By cor. 3.2.2 and cor. 3.2.4, it is enough to show that the projective line $\mathbf{P}_1$ over $\mathbf{Q}$ has the Hilbert property. There are several ways of proving this:

1. Hilbert's original method ([Hi]): The proof uses Puiseux expansions (cf. [L]). It shows that if $A \subset \mathbf{A}^1(\mathbf{Q})$ is thin, then the number of integers $n \in \mathbf{Z} \cap A$ with $n < N$ is $O(N^\delta)$ for some $\delta < 1$, when $N \to \infty$. (But it does not give a good estimate for δ.)

2. Proof by counting points of height smaller than N on $\mathbf{P}_1(\mathbf{Q})$: write $\xi \in \mathbf{P}_1(\mathbf{Q})$ as (x, y) with $x, y \in \mathbf{Z}$ and x, y relatively prime. This can be done in a unique way, up to a choice of sign. The *height* of ξ is defined to be
$$\text{height}(\xi) = \sup(|x|, |y|).$$
The number of points in $\mathbf{P}_1(\mathbf{Q})$ with height less than N is asymptotically $\frac{12}{\pi^2} N^2$. On the other hand:

Proposition 3.4.2 *If $A \subset \mathbf{P}_1(\mathbf{Q})$ is thin, then the number of points of A with height $\leq N$ is $\ll N$, when $N \longrightarrow \infty$.*

Sketch of proof: We may assume A is of type (C_2), $A \subset \pi(X(\mathbf{Q}))$, where X is is an absolutely irreducible curve and $\pi : X \longrightarrow \mathbf{P}_1$ has degree ≥ 2. Let $\text{Rat}_X(N)$ be the number of points x in $X(\mathbf{Q})$ with $\text{height}(\pi(x)) \leq N$, and let g be the genus of X.
Case 1: $g \geq 2$. Then $X(\mathbf{Q})$ is finite by Faltings's theorem; hence $\text{Rat}_X(N)$ is $O(1)$. One could also invoke an earlier result of Mumford (cf. [L], [Se9]) which gives $\text{Rat}_X(N) = O(\log \log N)$.
Case 2: $g = 1$: By the Mordell-Weil theorem, and the Néron-Tate theory of normalized heights, (see [L]), one has $\text{Rat}_X(N) = O((\log N)^{\gamma/2})$, where γ is the rank of $X(\mathbf{Q})$.
Case 3: $g = 0$: let H_X, H denote the heights on X and $\mathbf{P}_1$ respectively. It is known that
$$H_X \stackrel{\smile}{\frown} (H \circ \pi)^m,$$
where $m = \deg \pi$. Hence, the number of points on X with $H \circ \pi \leq N$ is at most $O(N^{1/m})$.

3. Proof by counting integral points: A variant of the second proof shows that the number of integral points in A with height less than N is $O(N^{\frac{1}{2}})$, which is an optimal bound.

4. Proof by counting S-units: let $S = \{p_1, \ldots, p_k\}$ be a finite non-empty set of primes, $E_S = \{\pm p_1^{m_1} \cdots p_k^{m_k}\}, m_i \in \mathbf{Z}$. Let $\alpha \in \mathbf{Q}$, and let $\alpha + E_S$ be the set of $\alpha + e$, where $e \in E_S$.

Proposition 3.4.3 *If A is a thin set in $\mathbf{P}_1(\mathbf{Q})$, then $A \cap (\alpha + E_S)$ is finite for all but a finite number of α.*

Recall that, if V is an affine variety over $\mathbf{Q}$, a subset B of $V(\mathbf{Q})$ is called quasi-S-integral if for every regular function f on V, the set $f(B)$ has bounded denominators in the ring of S-integers, i.e., there is a non-zero integer θ (depending on f and B) such that $\theta f(b)$ is an S-integer for all $b \in B$.

Let now $\pi_j : X_j \longrightarrow \mathbf{P}_1$ be a finite number of coverings, and choose α outside the finite sets $\mathrm{ram}(\pi_j)$ of X at which π_j is ramified. If $B_j \subset X_j(\mathbf{Q})$ is the set of elements of $X_j(\mathbf{Q})$ with $\pi_j(b) \in \alpha + E_S$, then B_j is a quasi-S-integral set in the affine curve $X_j - \pi_j^{-1}(\infty) - \pi_j^{-1}(\alpha)$. Over $\bar{\mathbf{Q}}$, $\pi_j^{-1}(\infty) \cup \pi_j^{-1}(\alpha)$ has at least 3 elements, since $\pi_j^{-1}(\alpha) \notin \mathrm{ram}(\pi_j)$. A theorem of Siegel, Mahler and Lang then shows that B_j is finite ([L], [Se9]). Prop. 3.4.3 follows.

Remark: The bound given in proof no. 3 can be extended to affine n-space $\mathbf{A}^n$. More precisely, let A be a thin set in $\mathbf{A}^n(\mathbf{Q})$, and let $\mathrm{Int}_A(N)$ be the number of integral points $(x_1, \ldots, x_n) \in A$ with $|x_i| \leq N$. Then:

Theorem 3.4.4 (S.D. Cohen) $\mathrm{Int}_A(N) = O(N^{n-\frac{1}{2}} \log N)$.

One can replace the $\log N$ term in this inequality by $(\log N)^\gamma$, where $\gamma < 1$ is a constant depending on A. The proof is based on the large sieve inequality: one combines th. 3.6.2 with cor. 10.1.2 of the appendix, cf. [Coh] and [Se9].

Problem: Let $X \subset \mathbf{P}_n$ be an absolutely irreducible variety of dimension r. As above, denote by $\mathrm{Rat}_X(N)$ the number of points of $X(\mathbf{Q})$ with height $\leq N$. If X is linear, $\deg X = 1$, then

$$\mathrm{Rat}_X(N) \asymp N^{r+1}.$$

If $\deg X \geq 2$, one can show, using th. 3.4.4 (cf. [Se9]), that

$$\mathrm{Rat}_X(N) = O(N^{r+\frac{1}{2}} \log N).$$

A better result follows from results of Schmidt [Schm], namely,

$$\mathrm{Rat}_X(N) = O(N^{r+\frac{4}{9}}).$$

Can this estimate be improved to:

$$\mathrm{Rat}_X(N) = O(N^{r+\epsilon}), \text{ for every } \epsilon > 0?$$

3.5 Hilbert property and weak approximation

Let K be a number field, Σ_K the set of places in K (including the archimedean ones). For $v \in \Sigma_K$, let K_v denote the completion of K at v, and let Nv be the cardinality of the residue field of K_v in case v is non-archimedean. If V is an absolutely irreducible integral variety over K, $V(K_v)$ is naturally endowed with a K_v-topology which gives it the structure of a K_v-analytic space (resp. manifold, if V is smooth).

Proposition 3.5.1 *If $W \subset V$, $W \neq V$, then $W(K_v) \neq V(K_v)$ for all but a finite number of $v \in \Sigma_K$.*

Proposition 3.5.2 *Let W be absolutely irreducible, of same dimension as V, and $\pi : W \longrightarrow V$ be a generically surjective morphism, $\deg \pi > 1$. Let K_π be the algebraic closure of K in the extension $K(W)^{\mathrm{gal}}/K(V)$. If $v \in \Sigma_K$ splits completely in K_π and Nv is large enough, then $\pi(W(K_v)) \neq V(K_v)$.*

The proofs of prop. 3.5.1 and 3.5.2 will be given in § 3.6.

Example: Take $K = \mathbf{Q}$, $V = \mathbf{P}_1$, W the curve defined by

$$y^3 = (x^2 + 3)/(x^2 + 12),$$

and define $\pi : W \longrightarrow V$ by $\pi(x, y) = x$. Then $K_\pi = \mathbf{Q}(\sqrt{-3})$. A prime p splits in K_π when $p \equiv 1 \pmod 3$. If $p \equiv -1 \pmod 3$, $p \geq 5$, then one checks that $\pi : W(\mathbf{Q}_p) \longrightarrow V(\mathbf{Q}_p)$ is an isomorphism of analytic $\mathbf{Q}_p$-manifolds. This shows that the condition "v splits completely in K_π" cannot be omitted.

Theorem 3.5.3 *Let A be a thin subset of $V(K)$, and let S_0 be a finite subset of Σ_K; then there is a finite set S of places of K satisfying:*
  a) $S \cap S_0 = \emptyset$.
  b) *The image of A in $\prod_{v \in S} V(K_v)$ is not dense.*

Observe first that if the theorem holds for A_1 and A_2, it holds also for $A_1 \cup A_2$: for, choose S_1 satisfying the conclusion of the theorem for A_1; then choose S_2 satisfying the conclusion of the theorem for A_2, but with S_0 replaced by $S_0 \cup S_1$. Then, taking $S = S_1 \cup S_2$, one has

$$A_1 \cup A_2 \subset \prod_{v \in S} V(K_v)$$

is not dense: the point (x_1, x_2), where $x_i \in \prod_{v \in S_i} V(K_v) - \text{closure}(A_i)$, is not in the closure of $A_1 \cup A_2$ in $\prod_{v \in S} V(K_v)$.

Hence it is enough to prove th. 3.5.3 for sets of type (C_1) and (C_2).

1. If $W \subset V$, $W \neq V$, $A \subset W(K)$, then choose $S = \{v\}$ with v large enough, so that $v \notin S_0$, $W(K_v) \neq V(K_v)$ (prop. 3.5.1). Since $\bar{A} \subset W(K_v)$, A is not dense in $W(K_v)$.

2. Assume $A \subset \pi(W(K))$, where $\pi : W \to V$ is generically surjective, $\dim W = \dim V$, W is absolutely irreducible, and $\deg \pi \geq 2$. We may also assume that π is a finite morphism (replace V by a suitable open subvariety). By prop. 3.5.2, there exists a $v \notin S_0$ such that $\pi(W(K_v)) \neq V(K_v)$. Since π is finite, it transforms closed subsets into closed subsets. This shows that $\pi(W(K_v))$ is closed in $V(K_v)$; hence A is not dense in $V(K_v)$.

Corollary 3.5.4 *Assume V is a projective variety, and let $\bar{A}$ denote the closure of A in the compact space*

$$\prod_{s \notin S_0} V(K_v).$$

Then the interior of $\bar{A}$ is empty, i.e., A is nowhere dense in the product.

One says that V has the *weak approximation property for a finite set S of places* if $V(K)$ is dense in $\prod_{v \in S} V(K_v)$.

Lemma 3.5.5 *If V, V' are smooth, birationally equivalent, then V has the weak approximation property for S if and only if V' has the weak approximation property for S. (In other words, the weak approximation property is a birational property for smooth varieties).*

It is enough to prove the lemma when $V' = V - W$, with W a closed subvariety, $W \neq V$. Clearly, if V has the weak approximation property for S, so does $V - W$. Conversely, if $V - W$ has the weak approximation property for S, one uses smoothness to prove that $V(K_v) - W(K_v)$ is dense in $V(K_v)$. Hence, any point in $W(K_v)$ can be approximated by points in $V(K) - W(K)$.

As a special case, any smooth K-rational variety has the weak approximation property for any finite set S of places.

Remark: The smoothness assumption is necessary: for example, consider the $\mathbf{Q}$-rational curve
$$y^2 = (x^2 - 5)^2 (2 - x^2).$$

Its rational points are not dense in the set of its real points (the points $(-\sqrt{5},0)$ and $(\sqrt{5},0)$ are isolated).

Definition 3.5.6 *A variety V is said to have* property (WA) *if it satisfies the weak approximation property with respect to S for all finite $S \subset \Sigma_K$. It is said to have* property (WWA) *("weak weak approximation property") if there exists a finite set S_0 of places of K such that V has the weak approximation property with respect to $S \subset \Sigma_K$, for all S with $S \cap S_0 = \emptyset$.*

Examples:
1. A K-rational variety has property (WA).
2. A K-torus has property (WWA), but not necessarily (WA). More precisely, if it is split by a finite Galois extension L/K, one can take for exceptional set S_0 the places of K whose decomposition group in $\mathrm{Gal}(L/K)$ is not cyclic. (See, e.g., [Vo1], [CTS1].)

Theorem 3.5.7 ([Ek], [CT]) *A variety which has the WWA property satisfies the Hilbert property.*

Proof: If not, $A = V(K)$ would be thin. By th. 3.5.3 there would exist S disjoint from S_0 such that $V(K)$ would not be dense in $\prod_{v \in S} V(K_v)$; this contradicts WWA.

The following conjecture is due to Colliot-Thélène [CT]; it is closely related to the questions discussed in [CTS2]:

Conjecture 3.5.8 *A K-unirational smooth variety has the WWA property.*

Recall that a variety is K-unirational if there exists a generically surjective map $\mathbf{P}_n \longrightarrow V$ defined over K, for some n - one may always take n equal to $\dim V$.

Theorem 3.5.9 *Conjecture 3.5.8 implies that every finite group is a Galois group over $\mathbf{Q}$.*

Proof: Make G act faithfully on $W = \mathbf{A}^n$ for some n, and let $V = W/G$. Then V is K-unirational. By conjecture 3.5.8, V^{smooth} has the WWA property, and hence satisfies the Hilbert property. By cor. 3.3.2, G can be realized as a Galois group over $\mathbf{Q}$.

3.6   Proofs of prop. 3.5.1 and 3.5.2

Let $\mathcal{O}$ denote the ring of integers of K, and choose a scheme $\underline{V}$ of finite type over $\mathcal{O}$ having V as its generic fiber; any two choices for $\underline{V}$ coincide outside a finite set of primes, i.e., they become isomorphic as schemes over $\mathrm{Spec}\mathcal{O}[\frac{1}{d}]$ for some non-zero d. If v is a non-archimedean place, $\mathcal{P}_v$ the corresponding prime of $\mathcal{O}$, denote by $\kappa(v) = \mathcal{O}/\mathcal{P}_v$ the residue field at v, which is a finite field with Nv elements. Let $\underline{V}(\kappa(v))$ be the set of $\kappa(v)$-rational points of $\underline{V}$ (or, equivalently, of the fiber of V at v). We shall use the following known result:

Theorem 3.6.1 (Lang-Weil) *If V is absolutely irreducible over K, then*

$$|\underline{V}(\kappa(v))| = Nv^{\dim V} + O((Nv)^{\dim V - \frac{1}{2}}).$$

The original proof of Lang-Weil is by reduction to the case of curves for which one can use the bound proved by Weil. A different method, which gives a more precise error term, is to use Deligne's estimates for the eigenvalues of the Frobenius endomorphism, together with Bombieri's bounds for the number of zeros and poles of the zeta function, see [Bo].

Proof of prop. 3.5.1
We may assume that V is smooth and $W = \emptyset$ (by replacing V by $V - W$). If Nv is large enough, th. 3.6.1 implies that $\underline{V}(\kappa(v))$ is not empty. Choose $x \in \underline{V}(\kappa(v))$. Since $\underline{V}$ is smooth at v (for Nv large enough), x lifts to a point $\tilde{x} \in \underline{V}(\mathcal{O}_v)$, which is contained in $V(K_v)$. Hence $V(K_v)$ is not empty.

Proof of prop. 3.5.2
Recall that we are given a generically surjective map $\pi : W \longrightarrow V$, with $\deg \pi > 1$, $\dim W = \dim V$. By replacing W and V by open subsets if necessary, we may assume that π is finite étale and V is smooth. We may also choose a scheme of finite type $\underline{W}$ for W over some $\mathcal{O}[\frac{1}{d}]$, such that π comes from a map (also denoted π) $\underline{W} \to \underline{V}$. By changing d, we may further assume that $\pi : \underline{W} \longrightarrow \underline{V}$ is étale and finite and $\underline{V}$ is smooth (see, e.g. EGA IV §8). For v prime to d, we have a diagram

$$
\begin{array}{ccccc}
\underline{W}(\kappa(v)) & \longleftarrow & \underline{W}(\mathcal{O}_v) & \hookrightarrow & W(K_v) \\
\downarrow & & \downarrow & & \downarrow \\
\underline{V}(\kappa(v)) & \longleftarrow & \underline{V}(\mathcal{O}_v) & \hookrightarrow & V(K_v).
\end{array}
$$

The fact that $\pi : \underline{W} \longrightarrow \underline{V}$ is finite implies that the right square is Cartesian, i.e. a point $z \in W(K_v)$ is an $\mathcal{O}_v$-point if and only if $\pi(z) \in V(K_v)$ has the same property. Moreover, the reduction map $\underline{V}(\mathcal{O}_v) \longrightarrow \underline{V}((v))$ is surjective, since $\underline{V}$ is smooth.

To show that $W(K_v) \neq V(K_v)$ it is thus enough to prove that

$$\underline{W}(\kappa(v)) \neq \underline{V}(\kappa(v)).$$

This is a consequence of the following more precise result:

Theorem 3.6.2 *Let $m = \deg \pi$ $(m \geq 2)$. Then, for v splitting completely in K_π, one has*

$$|\pi(\underline{W}(\kappa(v)))| \leq c(Nv)^{\dim V} + O((Nv)^{\dim V - \frac{1}{2}}),$$

where $c = 1 - \frac{1}{m!}$. (This implies:

$$|\pi(\underline{W}(\kappa(v)))| \leq c'(Nv)^{\dim V}$$

with $c' < 1$, for Nv large enough, v splitting completely in K_π.)

Let $\underline{W}^{\text{gal}}$ be the Galois closure of $\underline{W}$; its Galois group G injects into S_m, and hence $|G| \leq m!$. Now, divide the points in $\underline{W}(\kappa(v))$ into two sets:

$$\underline{W}(\kappa(v)) = A \cup B,$$

where A is the set of points which can be lifted to $\underline{W}^{\text{gal}}(\kappa(v))$, and B is the set of the remaining points. By th. 3.6.1 applied to $\underline{W}$, we have:

$$|A| + |B| = (Nv)^{\dim V} + O((Nv)^{\dim V - \frac{1}{2}}).$$

If v splits completely in K_π, then all the connected components of the fiber at v are absolutely irreducible, and hence, letting e be the number of these components, we have

$$|\underline{W}^{\text{gal}}(\kappa(v))| = e(Nv)^{\dim V} + O((Nv)^{\dim V - \frac{1}{2}}).$$

We have $A = \underline{W}^{\text{gal}}(\kappa(v))/H$ where $H = \text{Gal}(\underline{W}^{\text{gal}}/\underline{W})$. Since H acts freely, this gives

$$|A| = \frac{e}{|H|}(Nv)^{\dim V} + O((Nv)^{\dim V - \frac{1}{2}}).$$

The same argument applied to the action of G on $\underline{W}^{\text{gal}}(\kappa(v))$ gives

$$|\pi(A)| = \frac{e}{|G|}(Nv)^{\dim V} + O((Nv)^{\dim V - \frac{1}{2}})$$

$$= \frac{1}{m}|A| + O((Nv)^{\dim V - \frac{1}{2}}).$$

Hence:

$$|\pi(A)| + |\pi(B)| \quad \leq \quad \frac{1}{m}|A| + |B| + O((Nv)^{\dim V - \frac{1}{2}})$$

$$\leq \quad |A| + |B| - (1 - \frac{1}{m})|A| + O((Nv)^{\dim V - \frac{1}{2}})$$

$$\leq \quad (Nv)^{\dim V} - (1 - \frac{1}{m})|A| + O((Nv)^{\dim V - \frac{1}{2}}),$$

and therefore

$$\pi(\underline{W}(\kappa(v))) \leq (Nv)^{\dim V} - (1 - \frac{1}{m})\frac{e}{|H|}(Nv)^{\dim V} + O((Nv)^{\dim V - \frac{1}{2}}).$$

Finally, since $(1 - \frac{1}{m})\frac{e}{|H|} \geq \frac{1}{m|H|} \geq \frac{1}{|G|} \geq \frac{1}{m!}$, we get:

$$\pi(\underline{W}(\kappa(v))) \leq (1 - \frac{1}{m!})(Nv)^{\dim V} + O((Nv)^{\dim V - \frac{1}{2}}),$$

and this completes the proof of th. 3.6.2, and hence of prop. 3.5.2.

Application to the distribution of Frobenius elements

Let E be an elliptic curve over $\mathbf{Q}$ without complex multiplication over $\bar{\mathbf{Q}}$, and let a_p, for p prime, be the "trace of Frobenius":

$$a_p = 1 + p - N_p$$

where N_p is the number of points of $\underline{E}$ over $\mathbf{F}_p$. The following is well-known:

Theorem 3.6.3 *If $f \neq 0$ is any polynomial in two variables over $\mathbf{Q}$, then the set of primes p such that $f(p, a_p) = 0$ has density 0.*

(The proof uses the l-adic representation $\rho : G_{\mathbf{Q}} \longrightarrow \mathbf{GL}_2(\mathbf{Z}_l)$ attached to E: since $\det \rho(\mathrm{Frob}_p) = p$, $\mathrm{Tr}\, \rho(\mathrm{Frob}_p) = a_p$, one is lead to consider the set of $x \in \rho(G_{\mathbf{Q}})$ such that $f(\mathrm{Tr}\, x, \det x) = 0$. Since $\rho(G_{\mathbf{Q}})$ is known to be open in $\mathbf{GL}_2(\mathbf{Z}_l)$, this set has Haar measure zero; the theorem follows by applying Chebotarev's density theorem.)

More generally,

Theorem 3.6.4 *Let A be a thin subset of $\mathbf{Z} \times \mathbf{Z}$. The set of p's such that $(p, a_p) \in A$ has density 0.*

The theorem is already proved for A of type (C_1): so assume A is of type (C_2). Let A_l be the image of A in $\mathbf{Z}/l\mathbf{Z} \times \mathbf{Z}/l\mathbf{Z}$ and let S_l be the set of $x \in \mathbf{GL}_2(\mathbf{Z}/l\mathbf{Z})$ such that $(\mathrm{Tr}\,(x), \det(x))$ belongs to A_l. One checks that

$$|S_l| \leq |A_l| \cdot l^2(1 + 1/l).$$

By th. 3.6.2, we have $|A_l| \leq cl^2$ for $c < 1$ and l sufficiently large, splitting completely in some fixed extension K of $\mathbf{Q}$; hence

$$|S_l| \leq cl^4(1 + 1/l).$$

This shows that the density of S_l in $\mathbf{GL}_2(\mathbf{Z}/l\mathbf{Z})$ is

$$\leq c(1 + 1/l)(1 - 1/l)^{-1}(1 - 1/l^2)^{-1} = c(1 - 1/l)^{-2}$$

if l is large enough. Recall now that, if $l_1, \cdots, l_m$ are large enough distinct prime numbers, the Galois group of the $l_1 \cdots l_m$-division points of E is $\prod \mathbf{GL}_2(\mathbf{Z}/l_i\mathbf{Z})$. If each l_i splits completely in K the Chebotarev density theorem, applied to the field of $l_1 \cdots l_m$-division points of E, shows that the upper density of the set of primes p with $(p, a_p) \in A$ is $\leq c^m \prod_{i=1}^{m}(1 - 1/l_i)^{-2}$. Since this can be made arbitrarily small by taking m large enough, the theorem follows.

Remarks:
1. The above implies, for example, that the set of the primes p for which $|E(\mathbf{F}_p)| - 3$ is a square has density 0.
2. One can prove more than density zero in 3.6.4; in fact, one has

$$|\{p|p < N, (p, a_p) \in A\}| = O\left(\frac{N}{(\log N)^{1+\delta}}\right), \quad \text{for some } \delta > 0.$$

This implies that

$$\sum_{(p, a_p) \in A} \frac{1}{p} < \infty.$$

The proof (unpublished) uses the Selberg "Λ^2" sieve.
3. There are similar results for the Ramanujan τ-function.

Chapter 4

Galois extensions of Q(T): first examples

4.1 The property Gal$_T$

Let E be a finite Galois extension of $\mathbf{Q}(T)$ with group G which is *regular*, i.e., $\bar{\mathbf{Q}} \cap E = \mathbf{Q}$. Geometrically, E can be viewed as the function field of a smooth projective curve C which is absolutely irreducible over $\mathbf{Q}$; the inclusion $\mathbf{Q}(T) \hookrightarrow E$ corresponds to a (ramified) Galois covering $C \longrightarrow \mathbf{P}_1$ defined over $\mathbf{Q}$ with group G.

Conjecture 4.1.1 *Every finite group G occurs as the Galois group of such a covering.*

Let us say that G has property Gal$_T$ if there is a regular G-covering $C \longrightarrow \mathbf{P}_1$ as above. In that case, there are infinitely many linearly disjoint extensions of $\mathbf{Q}$, with Galois group G (cf. th. 3.3.3).
Remark: If a regular G-covering exists over $\mathbf{P}_n$, $n \geq 1$, then such a covering also exists over $\mathbf{P}_1$, by Bertini's theorem (cf. e.g. [Jou]).

Examples: The property Gal$_T$ is satisfied for:
1. Abelian groups.
2. A_n and S_n (Hilbert); $\tilde{A}_n$ (Vila, Mestre).
3. Some non-abelian simple groups, such as the sporadic ones (with the possible exception of M_{23}), most $\mathbf{PSL}_2(\mathbf{F}_p)$, p prime, and a few others.
4. If G has property Gal$_T$, then so does every quotient of G.

Proposition 4.1.2 *If G_1, G_2 have property* Gal$_T$*, then so does their product $G_1 \times G_2$.*

Proof: Let C_1, C_2 be regular coverings of $\mathbf{P}_1$ with groups G_1, G_2, and let Σ_1, Σ_2 be their ramification loci.

1. If $\Sigma_1 \cap \Sigma_2 = \emptyset$, then the extensions corresponding to C_1, C_2 are linearly disjoint, because $\mathbf{P}_1$ is algebraically simply connected (see §4.4 below). One can take for covering C the fibered product $C_1 \times_{\mathbf{P}_1} C_2$, which has function field $\mathbf{Q}(C_1) \otimes_{\mathbf{Q}(T)} \mathbf{Q}(C_2)$, and hence has $G_1 \times G_2$ as Galois group.

2. If $\Sigma_1 \cap \Sigma_2 \neq \emptyset$, one modifies the covering $C_1 \longrightarrow \mathbf{P}_1$ by composing it with an automorphism of $\mathbf{P}_1$ so that the new ramification locus is disjoint from Σ_2. One is thus reduced to case 1.

Remark: If G_1 and G_2 have property Gal_T, one can show (cf. e.g. [Ma3], p. 229, Zusatz 1) that the wreath product $G_1 \,\mathrm{Wr}\, G_2$ also has property Gal_T. This can be used to give an alternate proof of prop. 4.2.2 below.

Exercises:

1. Show that the profinite group $\mathbf{Z}_p$ is not the Galois group of any regular extension of $\mathbf{Q}(T)$. (Hence conjecture 4.1.1 does not extend to profinite groups, not even when they are p-adic Lie groups.)

2. Let G be a finite group having property Gal_T. Show that there exists a regular Galois extension L of $\mathbf{Q}(T)$, with Galois group G, such that:

(a) Every subextension of L distinct from $\mathbf{Q}(T)$ has genus ≥ 2.

(b) Every $\mathbf{Q}$-rational point P of $\mathbf{P}_1$ has property $\mathrm{Irr}(P)$ with respect to L. (Use a suitable base change $\mathbf{P}_1 \longrightarrow \mathbf{P}_1$, combined with Faltings's theorem.)

4.2 Abelian groups

A torus defined over $\mathbf{Q}$ is said to be a "permutation torus" if its character group has a $\mathbf{Z}$-basis which is stable under the action of $\mathrm{Gal}(\bar{\mathbf{Q}}/\mathbf{Q})$, or equivalently, if it can be expressed as a product of tori of the form $R_{K_i/\mathbf{Q}}\mathbf{G}_m$, where the K_i are finite extensions of $\mathbf{Q}$. A permutation torus is clearly rational over $\mathbf{Q}$.

Now, let A be a finite abelian group. The following proposition implies that A has property Gal_T:

Proposition 4.2.1 *There exists a torus S over $\mathbf{Q}$, and an embedding of A in $S(\mathbf{Q})$, such that the quotient $S' = S/A$ is a permutation torus. (In particular, S' is a $\mathbf{Q}$-rational variety.)*

The proof uses the functor Y which to a torus associates the $\mathbf{Z}$-dual of its character group. An exact sequence of the form $1 \longrightarrow A \longrightarrow S \longrightarrow S' \longrightarrow 1$ gives rise to the exact sequence

$$1 \longrightarrow Y(S) \longrightarrow Y(S') \longrightarrow \tilde{A} \longrightarrow 1,$$

where
$$\tilde{A} = \text{Ext}^1(\hat{A}, \mathbf{Z}) = \text{Hom}(\mu_n, A) = \text{Hom}(\hat{A}, \mathbf{Q}/\mathbf{Z}),$$

and $\hat{A}$ denotes as usual the Cartier dual $\text{Hom}(A, \mathbf{G}_m)$. Choose K finite Galois over $\mathbf{Q}$ such that the action of $G_\mathbf{Q}$ on $\tilde{A}$ factors through $\text{Gal}(K/\mathbf{Q})$, e.g. $K = \mathbf{Q}(\mu_n)$, where n is the exponent of A. Now express $\tilde{A}$ as a quotient of a free $\mathbf{Z}[\text{Gal}(K/\mathbf{Q})]$-module F, and let S' be a torus such that $Y(S') = F$; it follows that S' is a permutation torus, and there is an A-isogeny $S \longrightarrow S'$.

Proposition 4.2.2 *Let G be a finite group having property Gal_T, and let M be a finite abelian group with G-action. Then the semi-direct product $\tilde{G} = M \cdot G$ also has property Gal_T.*

We may assume without loss of generality that M is an induced G-module, $M = \bigoplus_{g \in G} gA$, so that
$$\tilde{G} = \underbrace{(A \times \cdots \times A)}_{|G| \text{ times}} \cdot G$$

is the *wreath product* of A and G. By prop. 4.2.1, there is an isogeny $S \longrightarrow S'$ defined over $\mathbf{Q}$, with S' a permutation torus and with kernel $A \subset S(\mathbf{Q})$. By hypothesis, there is a regular étale G-covering $C \longrightarrow U$ where U is a $\mathbf{Q}$-rational variety (e.g. an open subvariety of $\mathbf{P}_1$). The actions of A on S, and of G on C and on $S \times \ldots \times S$ give rise to a natural $\tilde{G}$-action on $X = S \times \ldots \times S \times C$. This action is free. Let $Y = X/\tilde{G}$. Prop. 4.2.2 then follows from the following lemma:

Lemma 4.2.3 *The variety $Y = X/\tilde{G}$ is $\mathbf{Q}$-rational.*

Define $X' = X/(A \times \ldots \times A) = S' \times \ldots \times S' \times C$. We have $Y = X'/G$. This shows that Y is the fiber space over U with fiber the torus $S' \times \ldots \times S'$, which is associated to the principal G-bundle $C \longrightarrow U$. We may thus view Y as a *torus over U*. In particular, the generic fiber Y_U of $Y \longrightarrow U$ is a torus over the function field $\mathbf{Q}(U)$ (this torus is obtained from $S' \times \ldots \times S'$ by a G-twisting, using the Galois extension $\mathbf{Q}(C)/\mathbf{Q}(U)$). Since S' is a permutation torus over $\mathbf{Q}$, Y_U is a permutation torus over $\mathbf{Q}(U)$. Hence the function field $\mathbf{Q}(Y)$ of Y_U is a pure transcendental extension of $\mathbf{Q}(U)$, which itself is a pure transcendental extension of $\mathbf{Q}$. The lemma follows. (One can also deduce lemma 4.2.3 from lemma 4.3.1 below.)

Exercise: Let H be a finite group generated by an abelian normal subgroup M and a subgroup G having property Gal_T. Show using prop. 4.2.2 that H has property Gal_T.

4.3   Example: the quaternion group Q_8

We need first:

Lemma 4.3.1 *Let* $Y \longrightarrow X$ *be an étale Galois covering with Galois group* G, *and* $G \longrightarrow \mathbf{GL}(W)$ *be a linear representation of* G, *where* W *is a finite-dimensional vector space. Let* E *be the associated fiber bundle with base* X *and fiber* W. *Then* E *is birationally isomorphic to* $X \times W$.

This follows from Hilbert's theorem 90: $H^1(K(X), \mathbf{GL}_n) = 0$, where n is $\dim W$, and $K(X)$ is the function field of X over the ground field (which we assume to be $\mathbf{Q}$).

[Alternate proof: Use descent theory to show that E is a vector bundle over X, hence is locally trivial.]

This lemma implies that $\mathbf{Q}(E) = \mathbf{Q}(X)(T_1, \ldots, T_n)$, hence that $\mathbf{Q}(E)$ and $\mathbf{Q}(X)$ are stably isomorphic. (Recall that two extensions k_1 and k_2 of a field k are stably isomorphic if there exist integers $n_1, n_2 \geq 0$ such that the extensions $k_1(T_1, \ldots, T_{n_1})$ and $k_2(T_1, \ldots, T_{n_2})$ are k-isomorphic.)

Application: Let G act linearly on vector spaces W_1 and W_2, the action on W_2 being faithful. Letting $n = \dim W_1$, we have:

4.3.2
$$\mathbf{Q}(W_1 \times W_2)/G \simeq \mathbf{Q}(W_2/G)(T_1, \ldots, T_n).$$

(In particular, $\mathbf{Q}(W_1 \times W_2)/G$ is stably isomorphic to $\mathbf{Q}(W_2/G)$.) This is a consequence of the lemma applied to $E = W_1 \times W_2$ and $X = W_2$.

Corollary 4.3.3 *If* G *acts faithfully on* W_1 *and* W_2, *then* $\mathbf{Q}(W_1/G)$ *and* $\mathbf{Q}(W_2/G)$ *are stably isomorphic.*

Let $R = \oplus W_i$ be a decomposition of the regular representation of G as a sum of $\mathbf{Q}$-irreducible ones. By the corollary, if one of the W_i is a faithful G-module, then $\mathbf{Q}(R/G)$ is stably isomorphic to $\mathbf{Q}(W_i/G)$. If $\mathbf{Q}(W_i/G)$ is a rational field, then so is $\mathbf{Q}(R/G)$, by 4.3.2.

Application to the quaternion group Q_8.   Let Q_8 be the quaternion group of order 8. The group algebra $\mathbf{Q}[Q_8]$ decomposes as

$$\mathbf{Q}[Q_8] = \mathbf{Q} \times \mathbf{Q} \times \mathbf{Q} \times \mathbf{Q} \times \mathbf{H},$$

where $\mathbf{H}$ denotes the standard field of quaternions (over $\mathbf{Q}$), and Q_8 acts on $\mathbf{H}$ by left multiplication. By the previous remark, the $\mathbf{Q}$-rationality of $\mathbf{Q}[Q_8]/Q_8$ is equivalent to the $\mathbf{Q}$-rationality of $\mathbf{H}^*/Q_8$, where $\mathbf{H}^*$ is the multiplicative group of $\mathbf{H}$ viewed as a 4-dimensional $\mathbf{Q}$-algebraic group.

The group Q_8 has a center $\{\pm 1\}$ of order 2, and $D = Q_8/\{\pm 1\}$ is a group of type $(2,2)$. On the other hand,

$$\mathbf{H}^*/\{\pm 1\} \simeq \mathbf{G}_\mathrm{m} \times \mathbf{SO}_3,$$

by the map

$$(N,\phi) : \mathbf{H}^* \longrightarrow \mathbf{G}_\mathrm{m} \times \mathbf{SO}_3,$$

where N is the reduced norm, and $\phi : \mathbf{H}^* \longrightarrow \mathbf{SO}_3$ maps a quaternion x to the rotation $y \mapsto xyx^{-1}$ (on the 3-dimensional vector space of quaternions of trace 0). Therefore,

$$\mathbf{H}^*/Q_8 = \mathbf{G}_\mathrm{m} \times \mathbf{SO}_3/D,$$

and it suffices to show that $\mathbf{SO}_3/D$ is a rational variety over $\mathbf{Q}$. But the group D is the stabilizer in $\mathbf{SO}_3$ of a flag in $\mathbf{A}^3$. Hence $\mathbf{SO}_3/D$ is isomorphic to an open subvariety of the flag variety of $\mathbf{A}^3$, which is rational over $\mathbf{Q}$. Noether's method therefore applies to Q_8; in particular Q_8 has property Gal_T. For explicit formulae, see [JY].

Exercise (L. Schneps) Show that every p-group of order p^3 has property Gal_T (use the exercise at the end of §4.2).

4.4 Symmetric groups

The symmetric group S_n acts on the affine space $\mathbf{A}^n$ with quotient $\mathbf{A}^n$ ("symmetric functions theorem"). This shows that S_n has property Gal_T. Let us give some explicit constructions of polynomials with S_n as Galois group. For example, consider a polynomial

$$f(X) = X^n + a_1 X^{n-1} + \cdots + a_n, \text{ with } a_i \in \mathbf{Q},$$

and put

$$f(X,T) = X^n + a_1 X^{n-1} + \cdots + a_n - T.$$

Theorem 4.4.1 (Hilbert [Hi]) *If f is a Morse function, then the splitting field of $f(X,T)$ over $\mathbf{Q}(T)$ is a regular extension with Galois group S_n.*

(The polynomial f is called a Morse function if:
  1. The zeros $\beta_1, \ldots, \beta_{n-1}$ of the derivative f' of f are simple.
  2. $f(\beta_i) \neq f(\beta_j)$ for $i \neq j$.)

We will need the following simple facts about the symmetric group S_n:

Lemma 4.4.2 *S_n is generated by transpositions.*

This is well-known: indeed S_n is generated by the transpositions

$$(12), (23), \ldots, (n-1, n).$$

Lemma 4.4.3 *Let G be a transitive subgroup of S_n which contains a transposition. Then the following are equivalent:*
1. *G contains an $(n-1)$-cycle.*
2. *G is doubly transitive.*
3. *$G = S_n$.*

If G contains an $(n-1)$-cycle, then the stabilizer of a point is transitive on the complement of the point, hence G is doubly transitive. If G is doubly transitive, then G contains all the transpositions in S_n, hence $G = S_n$ by lemma 4.4.2. That $3 \Rightarrow 1$ is obvious.

Lemma 4.4.4 (cf. [Hu], p.171) *Let G be a transitive subgroup of S_n which is generated by cycles of prime orders. Then:*
1. *G is primitive.*
2. *If G contains a transposition, then $G = S_n$.*
3. *If G contains a 3-cycle, then $G = A_n$ or S_n.*

Let $\{Y_1, \ldots, Y_k\}$, with $k > 1$, be a partition of $\{1, \ldots, n\}$ which is stable under G. Our assumptions imply that there is a cycle s of G, of prime order p, such that $Y_1 \neq sY_1$. Since no element of $Y_1 \cup sY_1 \cup \ldots \cup s^{p-1}Y_1$ is fixed by s, we have:

$$|Y_1| + |sY_1| + \ldots + |s^{p-1}Y_1| \leq p,$$

and hence $|Y_1| = 1$. This shows that $\{Y_1, \ldots, Y_k\}$ is the trivial partition of $\{1, \ldots, n\}$. Hence G is primitive. To show 2, let G' be the subgroup of G generated by the transpositions belonging to G. Since $G' \neq 1$, it is transitive (a non-trivial normal subgroup of a primitive group is transitive). For $\Omega \subset \{1, \ldots, n\}$, let us denote by S_Ω (resp. A_Ω) the symmetric (resp, alternating) group on Ω. Let $\Omega \subset \{1, \ldots, n\}$ be maximal with the property $S_\Omega \subset G'$, and suppose that $\Omega \neq \{1, \ldots, n\}$. By the transitivity of G', there exists $(xy) \in G'$ with $x \in \Omega$, $y \notin \Omega$. Hence $S_{\Omega \cup \{y\}} \subset G'$, contradicting the maximality of Ω. It follows that $\Omega = \{1, \ldots, n\}$ and hence $G = G' = S_n$, proving 2. The proof of 3 is similar, taking G' this time to be the subgroup of G generated by the 3-cycles belonging to G. The hypothesis implies that G' is non-trivial, and hence is transitive. Choose $\Omega \subset \{1, \ldots, n\}$ maximal with the property $A_\Omega \subset G'$. As before, if $\Omega \neq \{1, \ldots, n\}$, there is a 3-cycle (xyz) which does not stabilize Ω. There are two cases:
Case 1: $\{x, y, z\} \cap \Omega$ has two elements, say y and z. Then clearly $A_{\Omega \cup \{x\}} \subset G'$, contradicting the maximality assumption for Ω.

Case 2: $\{x, y, z\} \cap \Omega$ has 1 element, say x. Choose two elements $y', z' \in \Omega$ distinct from x; it is easy to see that (xyz) and $(xy'z')$ generate the alternating group A_5 on $\{x, y, z, y', z'\}$. In particular, the cycle $(xy'z)$ is in G; since this 3-cycle meets Ω in two elements, we are reduced to case 1, QED.

More generally, Jordan has shown that a primitive subgroup of S_n which contains a cycle of prime order $\leq n - 3$ is equal to A_n or S_n (see [Wi], p.39).

Th. 4.4.1 will be proved in the following more general form:

Theorem 4.4.5 *If K is any field of characteristic 0, or of characteristic p not dividing n, and $f(X) \in K[X]$ is Morse, then $\mathrm{Gal}(f(X) - T) = S_n$ over $K(T)$.*

Proof: We may assume K to be algebraically closed. The polynomial f can be viewed as a ramified covering of degree n

$$f: \quad \mathbf{P}_1 \quad \longrightarrow \quad \mathbf{P}_1$$
$$x \quad \mapsto \quad t = f(x).$$

The corresponding field extension is $K(X) \supset K(T)$; it is separable because $p \nmid n$. Let $G \subset S_n$ be the Galois group of the Galois closure of $K(X)$ over $K(T)$, i.e., the Galois group of the equation $f(X) - T = 0$.

The ramification points of the covering f are

$$X = \infty, f(\beta_1), \dots, f(\beta_{n-1}).$$

At $X = \infty$, the ramification is tame, and the inertia group is generated by an n-cycle. At the $f(\beta_i)$, the hypothesis on f implies that the inertia group is tame for $p \neq 2$ and wild for $p = 2$, and that (in both cases) it is generated by a transposition.

Hence the theorem is a consequence of the following proposition, combined with lemma 4.4.4.

Proposition 4.4.6 *Let $C \longrightarrow \mathbf{P}_1$ be a regular Galois covering with group G, tamely ramified at ∞. Then G is generated by the inertia subgroups of points outside ∞, and their conjugates.*

Proof: Let H be the normal subgroup generated by the inertia subgroups outside ∞. Then C/H is a G/H-covering of $\mathbf{P}_1$ which is tame at ∞, and unramified outside. The Riemann-Hurwitz formula implies that the genus of C/H is $\leq \frac{1}{2}(1 - |G/H|)$; hence $G = H$.

Corollary 4.4.7
  a) $\mathbf{P}_1$ *is algebraically simply connected.*
  b) *In characteristic 0, the affine line is algebraically simply connected.*

Remark: In characteristic $p > 0$, the affine line $\mathbf{A}^1$ is not simply connected (as shown e.g. by the Artin-Schreier equation $X^p - X = T$). If G is the Galois group of an unramified covering of $\mathbf{A}^1$, then prop. 4.4.6 implies that G is generated by its Sylow p-subgroups. There is a conjecture by Abhyankar [Ab1] that, conversely, every group G having this property occurs. This has now been proved by Raynaud [Ra2] and generalized by Harbater [Harb].

Example of an S_n-extension of ramification type $(n, n-1, 2)$

The above method for constructing polynomials over $\mathbf{Q}(T)$ with Galois group S_n gives us polynomials with ramification type $(n, 2, \dots, 2)$. For a different example, consider the polynomial

$$f(X) = X^n - X^{n-1},$$

so that

$$f(X, T) = X^n - X^{n-1} - T.$$

Then

$$f' = nX^{n-2}(X - \alpha), \qquad \alpha = 1 - \frac{1}{n},$$

and hence the ramification is given (in char. 0) by

$$\begin{cases} \text{at } \infty : & \text{cycle of order } n; \\ \text{at } 0 : & \text{cycle of order } n - 1; \\ \text{at } \alpha : & \text{a transposition.} \end{cases}$$

Hence $G = S_n$ by lemma 4.4.3. The polynomial f has ramification type $(n, n-1, 2)$.

Remarks:
1. Consider $f(X, T) = X^{p+1} - X^p - T$. This polynomial has S_{p+1} as Galois group in characteristic different from p (the proof is similar to the one above). In characteristic p, one can show that it has Galois group $\mathbf{PGL}_2(\mathbf{F}_p)$.
2. One might ask for an explicit polynomial f_n over $\mathbf{Q}$ such that f_n has Galois group S_n. Here is an example: $f_n(X) = X^n - X - 1$. Indeed, Selmer [Sel] has shown that f_n is irreducible over $\mathbf{Q}$. Assuming this, let us prove that f_n has Galois group S_n. We look at the primes p dividing the discriminant of f_n, i.e., those modulo which $f_n(X)$ has a multiple root. This happens if $f_n(X)$ and $f'_n(X) = nX^{n-1} - 1$ have a common root mod p. Substituting $X^{n-1} \equiv 1/n$ in the equation $f(X) \equiv 0$, one gets $X \equiv n/(1 - n)$. Hence there can be at most one double root mod p for each ramified prime p. This shows that the inertia subgroup at p is either

trivial, or is of order two, generated by a transposition. But $G = \text{Gal}(f)$ is generated by its inertia subgroups, because $\mathbf{Q}$ has no non-trivial unramified extension. By Selmer's result, G is transitive; we have just shown that G is generated by transpositions; hence $G = S_n$ by lemma 4.4.4.

Many more examples can be found in the literature. For instance, the "truncated exponential"

$$1 + x + \frac{x^2}{2} + \frac{x^3}{6} + \ldots + \frac{x^n}{n!}$$

has Galois group S_n when $n \not\equiv 0 \pmod 4$, and Galois group A_n otherwise (I. Schur).

Exercises:

1. Let Y_n be the subvariety of $\mathbf{P}_{n-1}$ defined by the homogeneous equations

$$X_1^i + \cdots + X_n^i = 0 \qquad \text{for } i = 1, 2, \ldots, n-2.$$

1.1. Y_n is an absolutely irreducible smooth curve, whose genus g_n is given by:

$$g_n = 1 + (n-2)! \frac{n^2 - 5n + 2}{4}.$$

(e.g. $g_3 = g_4 = 0$, $g_5 = 4$, $g_6 = 49$, $g_7 = 481, \ldots$)

1.2. The quotient of Y_n by S_n (acting by permutation of coordinates) is isomorphic to $\mathbf{P}_1$.

1.3. The Galois covering $Y_n \longrightarrow \mathbf{P}_1$ is the Galois closure of the degree n covering given by the polynomial $f(X) = X^n - X^{n-1}$.

2. In characteristic 11, show that the equation

$$X^{11} + 2X^9 + 3X^8 - T^8 = 0$$

is an unramified extension of $\mathbf{A}^1$ whose Galois closure has for Galois group the Mathieu group M_{11}. (Hint: reduce mod 11 the equation of [Ma4], after dividing the X-variable by $11^{1/4}$.)

4.5   The alternating group A_n

One exhibits the alternating group A_n as a Galois group over $\mathbf{Q}(T)$ by using the following lemma ("double group trick").

Lemma 4.5.1 *Let G be the Galois group of a regular extension $K/k(T)$, ramified at most at three places which are rational over k, and let H be a subgroup of G of index 2. Then the fixed field K_1 of H is rational. (In particular, if $k = \mathbf{Q}$, then H has property Gal_T.)*

Because of the conditions on the ramification, the curve corresponding to K_1 has genus zero, and has a k-rational point. The lemma follows.

For example, the polynomials with ramification type $(n, n - 1, 2)$ discussed in the previous section can be used to construct A_n-extensions of $\mathbf{Q}(T)$. More precisely, let us change variables, and put

$$h(X, T) = (n - 1)X^n - nX^{n-1} + T.$$

Then the discriminant of h (with respect to X) is

$$\Delta(h) = (-1)^{\frac{n(n-1)}{2}} n^n (n - 1)^{n-1} T^{n-2}(T - 1).$$

Up to square factors, we have

$$\Delta(h) \sim \begin{cases} (-1)^{\frac{n}{2}}(n - 1)(T - 1) & \text{if } n \text{ is even;} \\ (-1)^{\frac{n-1}{2}} nT(T - 1) & \text{if } n \text{ is odd.} \end{cases}$$

Hence the equation $D^2 = \Delta$ defines a rational curve. For example, if n is even, by replacing T by $1 + (-1)^{\frac{n}{2}}(n - 1)T^2$, we get the equation

$$(n - 1)X^n - nX^{n-1} + 1 + (-1)^{\frac{n}{2}}(n - 1)T^2 = 0,$$

which gives rise to a regular Galois extension of $\mathbf{Q}(T)$ with Galois group A_n.

Hilbert's original construction [Hi] was somewhat different. For the sake of simplicity, we reproduce it here only in the case where $n = 2m$ is even. Choose a polynomial

$$g(X) = nX \prod_{i=1}^{m-1}(X - \beta_i)^2,$$

with the β_i distinct and non-zero. Then, take $f(X)$ so that $df/dX = g$. Assume that the $f(\beta_i)$ are all distinct, and distinct from $f(0)$. Then f has ramification type $(n, 2, 3, 3, \ldots, 3)$. Hence its Galois group is S_n by lemma 4.4.4. But then the quadratic subfield fixed by A_n is only ramified at the two places ∞ and 0; hence it is a rational field.

Exercise: Show that the condition that the $f(\beta_i)$ are $\neq f(0)$ can be suppressed.

4.6   Finding good specializations of T

Let f be a polynomial over $\mathbf{Q}(T)$ with splitting field a regular G-extension of $\mathbf{Q}(T)$. Although Hilbert's irreducibility theorem guarantees that for "most" values of t, the specialized polynomial $f(X, t)$ will have Galois group G over $\mathbf{Q}$, it does not give a constructive method for finding , say, an infinite number of such t's.

For $p \notin S$, where S denotes a suitable finite set of primes, the equation $f(X, T) = 0$ can be reduced mod p. If p is large enough (see exercise below), then all conjugacy classes in G occur as Frobenius elements at t for some $t \in \mathbf{F}_p$. Letting $C_1, \ldots, C_h$ be the conjugacy classes of G, one can thus find distinct primes $p_1, \ldots, p_h$ and points $t_1, \ldots, t_h$, with $t_i \in \mathbf{F}_{p_i}$, such that $\mathrm{Frob}_{p_i}(f(X, t_i)) = C_i$. Specializing T to any $t \in \mathbf{Q}$ such that $t \equiv t_i \pmod{p_i}$ for all i gives a polynomial $f(X) = f(X, t)$ whose Galois group over $\mathbf{Q}$ intersects each of the conjugacy classes C_i, and hence is equal to G, by the following elementary result:

Lemma 4.6.1 (Jordan, [J2]) *Let G be a finite group, and H a subgroup of G which meets every conjugacy class of G. Then $H = G$.*

Indeed, if H is a subgroup of G, then the union of the conjugates of H in G has at most

$$1 + (G : H)(|H| - 1) = |G| - ((G : H) - 1)$$

elements.

Remark: Here is an arithmetic application of lemma 4.6.1:

(*)   Let $f(X)$ be an irreducible polynomial of degree $n > 1$ with integral coefficients. For each prime p, denote by $N(f, p)$ the number of solutions of $f(x) \equiv 0 \pmod{p}$. Then the set of p's with $N(f, p) = 0$ has a density which is > 0 (and even $\geq 1/n$).

The proof is a straightforward application of Chebotarev's density theorem, cf. [Se12].

Exercises:
1. Use lemma 4.6.1 to show that every finite division algebra is commutative (cf. Bourbaki, A.VIII, §11, no.1).
2. Show that lemma 4.6.1 can be reformulated as: "Every transitive subgroup of S_n, $n \geq 2$, contains a permutation without fixed point". (This is how the lemma is stated in Jordan, [J2].)
3. Let $\pi : X \longrightarrow Y$ be a G-covering of absolutely irreducible projective smooth curves over $\mathbf{F}_p$. Let N be the number of geometric points of X where π is ramified, and let g be the genus of X. Assume that $1 + p - 2g\sqrt{p} > N$. Show that, for every conjugacy class c in G, there is a point $t \in Y(\mathbf{F}_p)$ over which π is unramified, and whose Frobenius class in G is c. (Apply Weil's bound to the curve X, twisted by an element of c.)

Chapter 5

Galois extensions of Q(T) given by torsion on elliptic curves

5.1 Statement of Shih's theorem

Consider an elliptic curve E over $\mathbf{Q}(T)$ with j-invariant equal to T, e.g. the curve defined by the equation

$$y^2 + xy = x^3 - \frac{36}{T - 1728}x - \frac{1}{T - 1728}.$$

(Any other choice of E differs from this one by a quadratic twist only.) By adjoining to $\mathbf{Q}(T)$ the coordinates of the n-division points of E, one obtains a Galois extension K_n of $\mathbf{Q}(T)$ with $\mathrm{Gal}(K_n/\mathbf{Q}(T)) = \mathbf{GL}_2(\mathbf{Z}/n\mathbf{Z})$. More precisely, the Galois group of $\mathbf{C} \cdot K_n$ over $\mathbf{C}(T)$ is $\mathbf{SL}_2(\mathbf{Z}/n\mathbf{Z})$, and the homomorphism $G_{\mathbf{Q}(T)} \longrightarrow \mathbf{GL}_2(\mathbf{Z}/n\mathbf{Z}) \xrightarrow{\det} (\mathbf{Z}/n\mathbf{Z})^*$ is the cyclotomic character. Hence the extension K_n is not regular when $n > 2$: the algebraic closure of $\mathbf{Q}$ in K_n is $\mathbf{Q}(\mu_n)$. So the method does not give regular extensions of $\mathbf{Q}(T)$ with Galois group $\mathbf{PGL}_2(\mathbf{F}_p)$, nor $\mathbf{PSL}_2(\mathbf{F}_p)$. Nevertheless, K-y. Shih was able to obtain the following result [Shih1], [Shih2]:

Theorem 5.1.1 *There exists a regular extension of* $\mathbf{Q}(T)$ *with Galois group* $\mathbf{PSL}_2(\mathbf{F}_p)$ *if* $\left(\frac{2}{p}\right) = -1$, $\left(\frac{3}{p}\right) = -1$, *or* $\left(\frac{7}{p}\right) = -1$.

Shih's theorem will be proved in §5.3.

Remark: It is also known that $\mathbf{PSL}_2(\mathbf{F}_p)$ has property Gal_T when $\left(\frac{5}{p}\right) = -1$; this follows from [Ml2] combined with th. 5.1.1.

5.2 An auxiliary construction

Let E be an elliptic curve defined over a field k of characteristic 0. The p-torsion of E, denoted $E[p]$, is a two-dimensional $\mathbf{F}_p$-vector space; let us call $PE[p]$ the associated projective line $PE[p] \simeq \mathbf{P}_1(\mathbf{F}_p)$. The actions of the Galois group G_k on $E[p]$ and $PE[p]$ give rise to representations

$$\rho : G_k \longrightarrow \mathbf{GL}(E[p]) \simeq \mathbf{GL}_2(\mathbf{F}_p)$$

and

$$\bar{\rho} : G_k \longrightarrow \mathbf{PGL}(PE[p]) \simeq \mathbf{PGL}_2(\mathbf{F}_p).$$

The determinant gives a well-defined homomorphism $\mathbf{PGL}_2(\mathbf{F}_p) \longrightarrow \mathbf{F}_p^*/\mathbf{F}_p^{*2}$. The group $\mathbf{F}_p^*/\mathbf{F}_p^{*2}$ is of order 2 when p is odd (which we assume from now on). Hence the homomorphism $\epsilon_p = \det \circ \bar{\rho}$ is a quadratic Galois character,

$$\epsilon_p : G_K \longrightarrow \{\pm 1\}.$$

The Weil pairing gives a canonical identification of $\wedge^2(E[p])$ with μ_p, so that $\det \rho = \chi$, where χ is the pth cyclotomic character giving the action of G_k on the pth roots of unity. Therefore ϵ_p is the quadratic character associated to $k(\sqrt{p^*})$, where $p^* = (-1)^{\frac{p-1}{2}} p$.

Let K be a quadratic extension of k, and let σ be the involution in $\mathrm{Gal}(K/k)$. Let N be an integer ≥ 1 and E an elliptic curve defined over K such that E and E^σ are N-isogenous (i.e., there is a homomorphism $\phi : E \longrightarrow E^\sigma$ with cyclic kernel of order N). Assume for simplicity that E has no complex multiplication, so that there are only two N-isogenies, ϕ and $-\phi$, from E to E^σ.

If $p \nmid N$, the maps $\phi, -\phi$ induce isomorphisms (also denoted $\phi, -\phi$) from $E[p]$ to $E^\sigma[p]$. By passing to the projective lines, one gets an isomorphism $\phi : PE[p] \longrightarrow PE^\sigma[p]$ which is independent of the choice of the N-isogeny $E \longrightarrow E^\sigma$, and commutes with the action of G_K. We define a map $\rho_{E,N} : G_k \longrightarrow \mathbf{PGL}(PE[p])$ as follows:

1. If s belongs to the subgroup G_K of G_k, then $\rho_{E,N}(s)$ is defined via the natural action of G_K on $PE[p]$.

2. If $s \in G_k - G_K$, then the image of s in $G_{K/k}$ is σ. Hence s gives an isomorphism $PE[p] \longrightarrow PE^\sigma[p]$, and $\rho(s) : PE[p] \longrightarrow PE[p]$ is defined by composing this isomorphism with $\phi^{-1} : PE^\sigma[p] \longrightarrow PE[p]$.

(In other words, $\rho_{E,N}$ describes the action of G_k on the pairs of points (x, y) in $PE[p] \times PE^\sigma[p]$ which correspond to each other under the isogeny ϕ.) One checks easily that the map

$$\rho_{E,N} : G_k \longrightarrow \mathbf{PGL}(PE[p])$$

so defined is a homomorphism.

As before, the projective representation $\rho_{E,N}$ of G_k gives a character

$$\epsilon_{E,N} : G_k \longrightarrow \pm 1.$$

Let ϵ_K denote the character $G_k \longrightarrow \pm 1$ corresponding to the quadratic extension K/k. Recall that $\epsilon_p : G_k \longrightarrow \pm 1$ corresponds to $k(\sqrt{p^*})/k$.

Proposition 5.2.1 *We have*:

$$\epsilon_{E,N} = \begin{cases} \epsilon_p & \text{if } \left(\dfrac{N}{p}\right) = 1 \\ \epsilon_p \epsilon_K & \text{if } \left(\dfrac{N}{p}\right) = -1 \end{cases}$$

Corollary 5.2.2 *If $K = k(\sqrt{p^*})$, and $\left(\dfrac{N}{p}\right) = -1$, then $\epsilon_{E,N} = 1$, i.e., the image of $\rho_{E,N}$ is contained in $\mathbf{PSL}_2(\mathbf{F}_p)$.*

Proof of 5.2.1: If $s \in G_K$, then $\rho_{E,N}(s)$ acts on $\wedge^2 E[p] \simeq \mu_p$ by $\chi(s)$. Hence $\epsilon_{E,N}(s) = \left(\dfrac{\chi(s)}{p}\right) = \epsilon_p(s)$. If $s \notin G_K$, then using the fact that the homomorphism $\phi^{-1} : E^\sigma[p] \longrightarrow E[p]$ induces multiplication by N^{-1} on

$$\mu_p \simeq \wedge^2 (E^\sigma[p]) \simeq \wedge^2 (E[p])$$

(where the group operation on μ_p is written additively), one finds that

$$\det \rho_{E,N}(s) = N^{-1}\chi(s).$$

Hence $\epsilon_{E,N}(s) = \left(\dfrac{N}{p}\right)\epsilon_p(s)$. This completes the proof.

5.3 Proof of Shih's theorem

Let K_N denote the function field of the modular curve $X_0(N)$ of level N over $\mathbf{Q}$,

$$K_N = \mathbf{Q}[j_1, j_2]/F_N(j_1, j_2),$$

where F_N is the (normalized) polynomial relating the j-invariants of N-isogenous elliptic curves (cf. [F]). The Fricke involution W_N acts on K_N by interchanging j_1 and j_2. Let $X_0(N)^p$ be the twist of $X_0(N)$ by the quadratic character ϵ_p, using the involution W_N. The function field of $X_0(N)^p$ can be described as follows: let k_N be the field of invariants of W_N in K_N, i.e., the function field of $X_0(N)^+$. Then K_N and $k_N(\sqrt{p^*})$ are disjoint quadratic extensions of k_N. The third quadratic extension, k, of k_N contained in $K = K_N(\sqrt{p^*})$ is the function field of $X_0(N)^p$

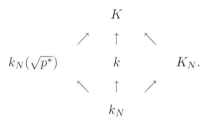

Now, let E denote an elliptic curve over K with j-invariant j_1. Then E^σ has j-invariant j_2, and hence E and E^σ are N-isogenous. Assuming $\left(\frac{N}{p}\right) = -1$, cor. 5.2.2 gives a projective representation $G_k \longrightarrow \mathbf{PSL}_2(\mathbf{F}_p)$. It is not hard to see that this representation is surjective, and gives a regular extension of k.

Remark: Instead of taking E over K one can take it over K_N. One thus gets a projective representation $G_{k_N} \longrightarrow \mathbf{PGL}_2(\mathbf{F}_p)$ and prop. 5.2.1 shows that it is surjective, and gives a regular extension if $(\frac{N}{p}) = -1$. Since k_N is known to be isomorphic to $\mathbf{Q}(T)$ if N belongs to the set

$$S = \{2, 3, 5, 7, 11, 13, 17, 19, 23, 29, 31, 41, 47, 59, 71\},$$

this shows that $\mathbf{PGL}_2(\mathbf{F}_p)$ has property Gal_T provided there exists $N \in S$ with $(\frac{N}{p}) = -1$ (a computer search shows that this is true for all $p < 5329271$).

We are interested in values of N for which k is isomorphic to $\mathbf{Q}(T)$, i.e., such that:

1. $X_0(N)$ has genus 0.
2. $X_0(N)^p$ has a $\mathbf{Q}$-rational point.

Assuming that N is not a square, condition 1 implies that $N = 2$, 3, 5, 6, 7, 8, 10, 12, 13, 18. Condition 2 is satisfied when $N = 2$, 3, 7 (and also for $N = 6$, $\left(\frac{2}{p}\right) = 1$ and $N = 10$, $\left(\frac{5}{p}\right) = 1$, cf. [Shih1]). More precisely:

Proposition 5.3.1 *For $N = 2$, 3, or 7, the two fixed points of W_N in $X_0(N)$ are rational over $\mathbf{Q}$.*

(Hence, these points stay rational on $X_0(N)^p$, which is thus isomorphic to $\mathbf{P}_1$ over $\mathbf{Q}$; this concludes the proof of th. 5.1.1.)

Let us give two proofs of prop. 5.3.1.

First proof: Let $\mathcal{O}$ be an order of class number 1 in a quadratic imaginary field, and $E_{\mathcal{O}}$ an elliptic curve with endomorphism ring $\mathcal{O}$. Assume that the unique prime ramified in the field is N, and that $\mathcal{O}$ contains an element π with $\pi\bar{\pi} = N$. This is indeed possible for $N = 2, 3, 7$:

N	$\mathcal{O}$	π
2	$\mathbf{Z}[i]$	$1+i$
	$\mathbf{Z}[\sqrt{-2}]$	$\sqrt{-2}$
3	$\mathbf{Z}\left[\frac{1+\sqrt{-3}}{2}\right]$	$\sqrt{-3}$
	$\mathbf{Z}[\sqrt{-3}]$	$\sqrt{-3}$
7	$\mathbf{Z}\left[\frac{1+\sqrt{-7}}{2}\right]$	$\sqrt{-7}$
	$\mathbf{Z}[\sqrt{-7}]$	$\sqrt{-7}$

By the theory of complex multiplication, the pair $(E_{\mathcal{O}}, \pi)$ then defines a rational point of $X_0(N)$ which is fixed under W_N.

Second proof: Let Δ be the discriminant modular form

$$\Delta(q) = q \prod (1 - q^n)^{24}, \qquad q = e^{2\pi i z},$$

and let f be the power series defined by

$$\begin{aligned} f &= (\Delta(z)/\Delta(Nz))^{\frac{1}{N-1}} \\ &= q^{-1} \prod_{\substack{n \not\equiv 0(N) \\ n \geq 1}} (1 - q^n)^{2m}, \qquad \text{where } m = \frac{12}{N-1}. \end{aligned}$$

One can prove that f generates K_N over $\mathbf{Q}$, and that $f^{W_N} = N^m/f$. For $N = 2, 3$, and 7 we have $m = 12, 6, 2$, and N^m is a square. This shows that the fixed points of W_N are rational.

One can also show that the quaternion algebra corresponding to $X_0(N)^p$ is given by (N^m, p^*). More generally:

Exercise: Let F be a field, and let X be the variety obtained by twisting $\mathbf{P}_1$ by the quadratic character attached to the extension $F(\sqrt{b})$ and the involution of $\mathbf{P}_1$ $x \mapsto a/x$. Then the quaternion invariant of this curve is the element (a, b) of $\mathrm{Br}_2(F)$.

The ramification in the Shih covering constructed above is as follows:

N=2: The ramification is of type $(2, p, p)$. One ramification point is rational over $\mathbf{Q}$, with inertia group of order 2 generated by the element $\begin{pmatrix} 0 & 1 \\ -1 & 0 \end{pmatrix}$. The two others are defined over $\mathbf{Q}(\sqrt{p^*})$ and are conjugate to each other. Their inertia groups, of order p, correspond to the two different conjugacy classes of unipotent elements in $\mathbf{PSL}_2(\mathbf{Z})$, $\begin{pmatrix} 1 & 1 \\ 0 & 1 \end{pmatrix}$ and $\begin{pmatrix} 1 & \alpha \\ 0 & 1 \end{pmatrix}$, where α is a non-quadratic residue mod p.

N=3: The situation is similar; we have ramification of type $(3, p, p)$, the inertia group of order 3 being generated by the element $\begin{pmatrix} 1 & -1 \\ 1 & 0 \end{pmatrix}$.

N=7: The ramification is of type $(3, 3, p, p)$; with a suitable choice of coordinates, the ramification points are located at $\pm\sqrt{-27p^*}$ and $\pm\sqrt{p^*}$.

Exercise: Show that the statement of th. 5.1.1 remains true when $\mathbf{PSL}_2(\mathbf{F}_p)$ is replaced by $\mathbf{PSL}_2(\mathbf{Z}/p^n\mathbf{Z})$.

5.4 A complement

Assume as before that $\left(\frac{N}{p}\right) = -1$. Then there is a regular Galois covering $C \longrightarrow X_0(N)^p$ with Galois group $G = \mathbf{PSL}_2(\mathbf{F}_p)$ by the above. In Shih's theorem, one takes N such that $X_0(N)$ has genus 0. One might try to exploit the case where $X_0(N)$ is of genus 1; for if the twisted curve $X_0(N)^p$ has a $\mathbf{Q}$-rational point, and is of rank > 0 (where $X_0(N)^p$ is viewed as an elliptic curve over $\mathbf{Q}$ by fixing this rational point as origin) then the following variant of Hilbert's irreducibility theorem allows us to deduce the existence of infinitely many extensions of $\mathbf{Q}$ with Galois group G:

Proposition 5.4.1 *Let* $C \longrightarrow E$ *be a regular Galois covering with group* G, *where* E *is an elliptic curve over a number field* K. *Assume that for every proper subgroup* H *of* G *containing the commutator subgroup* (G, G), *the corresponding* G/H-*covering of* E *is ramified at least at one point. Then all* $P \in E(K)$ *except a finite number have property* $\mathrm{Irr}(P)$ *of §3.3.*

The hypothesis in the proposition implies that for all subgroups $H \neq G$, the covering $C/H \longrightarrow E$ is ramified somewhere, since the only unramified coverings of E are abelian. Hence the genus of $C_H = C/H$ is at least 2; by Faltings' theorem, C_H has finitely many rational points. Let $S_K \subset E(K)$ be the union of the ramification points and the images of the $C_H(K)$. This is a finite set, and if $P \notin S_K$, then property $\mathrm{Irr}(P)$ is satisfied. This proves the proposition.

The above result was first obtained by Néron (Proc. Int. Cong. 1954) in a weaker form, since Mordell's conjecture was still unproved at that time.

Corollary 5.4.2 *Assume $E(K)$ is infinite. Then, there are infinitely many linearly disjoint Galois extensions of K with Galois group G.*

Proof: If L is any finite extension of K, we can find $P \in E(K)$, $P \notin S_L$ (where S_L is defined as above).The property Irr(P) is then satisfied both over K and over L. The corresponding G-extension K_P is then linearly disjoint from L. The corollary follows.

Example: For $N = 11$, $p = 47$, the twisted elliptic curve $X_0(N)^p$ has rank 2 over **Q**. Hence there are infinitely many extensions of **Q** with Galois group **PSL**$_2(\mathbf{F}_{47})$. (An explicit example has been written down by N. Elkies.)

Problem: Is it possible to generalize prop. 5.4.1 to abelian varieties? More precisely, let $\pi : C \longrightarrow A$ be a finite covering of an abelian variety A. Assume that a) $A(K)$ is Zariski dense in A; b) π is ramified (i.e., C is not an abelian variety). Is it true that $\pi(C(K))$ is "much smaller" than $A(K)$, i.e., that the number of points of logarithmic height $\leq N$ in $\pi(C(K))$ is $o(N^{\rho/2})$, where $\rho =$ rank $A(K)$? There is a partial result in this direction in the paper of A. Néron referred to above.

5.5 Further results on PSL$_2(\mathbf{F_q})$ and SL$_2(\mathbf{F_q})$ as Galois groups

Concerning the groups **PSL**$_2$ and **SL**$_2$ over finite fields, there are the following results:

1. **PSL**$_2(\mathbf{F}_q)$, for q not a prime, is known to have property Gal$_T$ when:

    (a) $q = 4$ and $q = 9$, because **PSL**$_2(\mathbf{F}_q)$ is isomorphic to A_5 and A_6 respectively in these cases.

    (b) $q = 8$, by a result of Matzat ([Ma3]).

    (c) $q = 25$, by a result of Pryzwara ([Pr]).

    (d) $q = p^2$, for p prime, $p \equiv \pm 2 \pmod 5$, cf. [Me1].

2. **SL**$_2(\mathbf{F}_q)$ is known to have property Gal$_T$ for:

    (a) $q = 2, 4, 8$, for then **SL**$_2(\mathbf{F}_q) \approx$ **PSL**$_2(\mathbf{F}_q)$.

    (b) $q = 3, 5$ or 9, for then **SL**$_2(\mathbf{F}_q)$ is isomorphic to $\tilde{A}_4$, $\tilde{A}_5$, or $\tilde{A}_6$, which are known to have property Gal$_T$, cf. [Me2], or §9.3.

    (c) $q = 7$, Mestre (unpublished).

It seems that the other values of q have not been treated (not even the case $q = 11$).

There are a few examples of Galois extensions of $\mathbf{Q}$ with Galois group $\mathbf{SL}_2(\mathbf{F}_{2^m})$ for $m = 1, 2, \ldots, 16$. Their construction is due to Mestre (unpublished), who uses the representations of $G_{\mathbf{Q}}$ given by modular forms mod 2 of prime level ≤ 600.

Chapter 6

Galois extensions of C(T)

6.1 The GAGA principle

Our goal in this chapter is to construct Galois extensions of $\mathbf{C}(T)$, using the tools of topology and analytic geometry. To carry out this program, we need a bridge between analysis and algebra.

Theorem 6.1.1 (GAGA principle) *Let X, Y be projective algebraic varieties over $\mathbf{C}$, and let X^{an}, Y^{an} be the corresponding complex analytic spaces. Then*
1. Every analytic map $X^{\mathrm{an}} \longrightarrow Y^{\mathrm{an}}$ is algebraic.
2. Every coherent analytic sheaf over X^{an} is algebraic, and its algebraic cohomology coincides with its analytic one.

For a proof, see [Se4] or [SGA1], exposé XII. In what follows, we will allow ourselves to write X instead of X^{an}.

Remarks:
1. The functor $X \mapsto X^{\mathrm{an}}$ is the "forgetful" functor which embeds the category of complex projective varieties into the category of complex analytic spaces. Th. 6.1.1 implies that it is fully faithful.
2. By the above, there is at most one algebraic structure on a compact analytic space which is compatible with it.
3. Th. 6.1.1 implies Chow's theorem: every closed analytic subspace of a projective algebraic variety is algebraic.
4. The analytic map $\exp : \mathbf{G}_{\mathrm{a}} \longrightarrow \mathbf{G}_{\mathrm{m}}$, where

$$\mathbf{G}_{\mathrm{a}} = \mathbf{P}_1 - \{\infty\}, \text{ and } \mathbf{G}_{\mathrm{m}} = \mathbf{P}_1 - \{0, \infty\},$$

is not algebraic; hence the hypothesis that X is projective is essential.

Exercise: If X and Y are reduced varieties of dimension 1, prove that any analytic isomorphism of X on Y is algebraic; disprove this for non-reduced varieties.

Theorem 6.1.2 (Riemann) *Any compact complex analytic manifold of dimension 1 is algebraic.*

The proof is easy, once one knows the finiteness of the (analytic) cohomology groups of coherent sheaves, see e.g. [Fo], chap. 2. A generalization for a broader class of varieties is given by a theorem of Kodaira:

Theorem 6.1.3 (see e.g. [GH], ch. I, §4) *Every compact Kähler manifold X whose Kähler class is integral (as an element of $H^2(X, \mathbf{R})$) is a projective algebraic variety.*

In the above theorems, the compactness assumption is essential. For coverings, no such assumption is necessary:

Theorem 6.1.4 *Let X be an algebraic variety over $\mathbf{C}$, and let $\pi : Y \longrightarrow X$ be a finite unramified analytic covering of X. Then there is a unique algebraic structure on Y compatible with its analytic structure and with π.*

The proof is given in [SGA1], exposé XII. Let us explain the idea in the case $X = \bar{X} - Z$, where $\bar{X}$ is an irreducible projective normal variety, and Z a closed subspace. One first uses a theorem of Grauert and Remmert [GR] to extend $Y \longrightarrow X$ to a ramified analytic covering $\bar{Y} \longrightarrow \bar{X}$. Such a covering corresponds to a coherent analytic sheaf of algebras over $\bar{X}$. Since $\bar{X}$ is projective, one can apply th. 6.1.1, and one then finds that this sheaf is algebraic, hence so are $\bar{Y}$ and Y.

(The extension theorem of Grauert and Remmert used in the proof is a rather delicate one - however, it is quite easy to prove when $\dim X = 1$, which is the only case we shall need.)

The GAGA principle applies to *real* projective algebraic varieties, in the following way: we may associate to any such variety X the pair (X^{an}, s), where $X^{\mathrm{an}} = X(\mathbf{C})$ is the complex analytic space underlying X, and

$$s : X^{\mathrm{an}} \longrightarrow X^{\mathrm{an}}$$

is the anti-holomorphic involution given by complex conjugation on X. The real variety X can be recovered from the data (X^{an}, s) up to a unique isomorphism. Furthermore, any complex projective variety X together with an anti-holomorphic involution s determines a projective variety over $\mathbf{R}$: for by GAGA, s is an algebraic isomorphism from X to $\bar{X}$, the conjugate variety of X, and hence gives descent data which determines X as a variety over $\mathbf{R}$. Similar remarks apply to coherent sheaves, and coverings: for example, giving a finite covering of X over $\mathbf{R}$ is equivalent to giving a finite covering of the complex analytic space $X(\mathbf{C})$, together with an anti-holomorphic involution which is compatible with the complex conjugation on $X(\mathbf{C})$.

6.2 Coverings of Riemann surfaces

Recall that for all $g \geq 0$, there exists, up to homeomorphism, a unique compact, connected, oriented surface X_g of genus g (i.e., with first Betti number $2g$). This can be proved either by "cutting and pasting" or by Morse theory (in the differentiable category). The surface X_g has a standard description as a polygon with $4g$ edges labelled a_1, b_1, a_1^{-1}, b_1^{-1}, ..., a_g, b_g, a_g^{-1}, b_g^{-1}, and identified in the appropriate manner. From this description one can see that the fundamental group $\pi_1(X_g)$ relative to any base point has a presentation given by $2g$ generators a_1, b_1, ..., a_g, b_g, and a single relation

$$a_1 b_1 a_1^{-1} b_1^{-1} \cdots a_g b_g a_g^{-1} b_g^{-1} = 1.$$

(To show this, one may express X_g as the union of two open sets U and V, where U is the polygon punctured at one point, which is homotopic to a wedge of $2g$ circles and has fundamental group the free group F on $2g$ generators, and V is a disk in the center of the polygon. Applying the van Kampen theorem, one finds that

$$\pi_1(X_g) = 1 *_R F = F/\langle R \rangle,$$

where R is $a_1 b_1 a_1^{-1} b_1^{-1} \ldots a_g b_g a_g^{-1} b_g^{-1}$.)

Let now $P_1, \ldots, P_k$ be distinct points of X_g, and let π_1 be the fundamental group of $X_g - \{P_1, \ldots, P_k\}$, relative to a base point x. Each P_i defines in π_1 a conjugacy class C_i corresponding to "turning around P_i in the positive direction" (choose a disk D_i containing P_i and no other P_j, and use the fact that $\pi_1(D_i - P_i)$ can be identified with $\mathbf{Z}$). A similar argument as above proves that the group π_1 has a presentation given by $2g + k$ generators a_1, b_1, ..., a_g, b_g, c_1, ..., c_k and the single relation:

$$a_1 b_1 a_1^{-1} b_1^{-1} \cdots a_g b_g a_g^{-1} b_g^{-1} c_1 \cdots c_k = 1,$$

where the c_i belong to C_i for all i.

We denote this group by $\pi_1(g, k)$. Observe that, if $k \geq 1$, then $\pi_1(g, k)$ is the free group on $2g + k - 1$ generators. In the applications, we shall be mainly interested in the case $g = 0$, $k \geq 3$.

6.3 From C to $\bar{\mathbf{Q}}$

Let K be an algebraically closed field of characteristic 0, X a projective smooth curve of genus g over K, and let P_1, ..., P_k be distinct points in $X(K)$. The algebraic fundamental group of $X - \{P_1, \ldots, P_k\}$ may be defined as follows. Let $\overline{K(X)}$ be an algebraic closure of the function field $K(X)$ of X over K, and let $\Omega \subset \overline{K(X)}$ be the maximal extension of $K(X)$

unramified outside the points $P_1, \ldots, P_k$. The algebraic fundamental group is the Galois group of Ω over $K(X)$, namely,

$$\pi_1^{\mathrm{alg}}(X - \{P_1, \ldots, P_k\}) = \mathrm{Gal}(\Omega/K(X)).$$

It is a profinite group. Let us denote it by π.

Theorem 6.3.1 *The algebraic fundamental group π is isomorphic to the profinite completion of the topological fundamental group:*

$$\pi \simeq \hat{\pi}_1(g, k).$$

(Recall that the profinite completion $\hat{\Gamma}$ of a discrete group Γ is the inverse limit of the finite quotients of Γ.)

Proof: Let first E be a finite extension of $K(X)$ with Galois group G, let Q_i be a place of E above P_i, and let $I(Q_i)$ denote the inertia subgroup of G at Q_i. Since the ramification is tame, we have a canonical isomorphism:

$$I(Q_i) \simeq \mu_{e_i},$$

where e_i is the ramification index at Q_i; this isomorphism sends an element s of $I(Q_i)$ to $s\pi/\pi$ (mod π), where π is any uniformizing element at Q_i.

From now on, let Q_i be a place of Ω above P_i. The isomorphism above can be generalized to such infinite extensions, by passing to the limit: write Ω as a union of finite extensions E_j with ramification indices e_j at Q_i. Then one has:

$$I(Q_i) \simeq \varprojlim \mu_{e_j}.$$

Choose a coherent system of roots of unity in K, i.e., a fixed generator z_n for each μ_n, such that $z_{nm}^m = z_n$ for all n, m. This provides $I(Q_i)$ with a canonical generator $c(Q_i)$. Changing Q_i above P_i merely affects $c(Q_i)$ by inner conjugation. Hence, the conjugacy class C_i of $c(Q_i)$ depends only on P_i. (Changing the coherent system $\{z_n\}$ – i.e., replacing it by $\{z_n^\alpha\}$, where $\alpha \in \hat{\mathbf{Z}}^* = \mathrm{Aut}\,\hat{\mathbf{Z}}$– replaces $c(Q_i)$ by $c(Q_i)^\alpha$.)

Th. 6.3.1 is a consequence of the following more precise statement:

Theorem 6.3.2 *There exists a choice of elements x_j, y_j, c_i in π, with $c_i \in C_i$ for all i, such that:*

1. $x_1 y_1 x_1^{-1} y_1^{-1} \cdots x_g y_g x_g^{-1} y_g^{-1} c_1 \cdots c_k = 1$.
2. The map $\pi_1(g, k) \longrightarrow \pi$ mapping the generators x_j, y_j, c_i of $\pi_1(g, k)$ to the elements x_j, y_i, c_i extends to an isomorphism

$$\hat{\pi}_1(g, k) \xrightarrow{\sim} \pi.$$

In other words, π is presented as a profinite group by the generators x_j, y_j, c_i with the relations above. The proof will be done in two stages: first, for $K = \mathbf{C}$, then, for arbitrary algebraically closed fields K of characteristic 0.

Step 1: Proof when $K = \mathbf{C}$, with the standard choice of roots of unity, $\{z_n\} = \{e^{2\pi i/n}\}$. In this case, the result follows from the GAGA dictionary between algebraic and topological coverings (see th. 6.1.3 and [SGA1]).

Step 2: K arbitrary. One has the following general result

Theorem 6.3.3 *Let V be an algebraic variety over an algebraically closed field K of characteristic 0, and let K' be an algebraically closed extension of K. Then any covering of V over K' comes uniquely from a covering defined over K.*

(Note that this is not true in characteristic p, unless the coverings are tame. For example, the Artin-Schreier covering $Y^p - Y = \alpha T$ of the affine line with $\alpha \in K'$, $\alpha \notin K$, does not come from a covering defined over K.)

Two coverings of V defined over K and K'-isomorphic are K-isomorphic; this is clear, e.g. by using a specialization argument. Next, one has to show that any covering which is defined over K' can be defined over K. If K' is of finite transcendence degree over K, then by using induction on tr.deg(K'/K), it is enough to show this for tr.deg$(K'/K) = 1$. In this case, a covering $W \longrightarrow V$ over K' corresponds to a covering $W \times C \longrightarrow V \times C$ over K, where C is the curve over K corresponding to the extension K'/K. This can be viewed as a family of coverings $W \longrightarrow V$ over K parameterized by C. But it is a general fact that in characteristic 0, such families of coverings are constant. (One can deduce this from the similar geometrical statement over $\mathbf{C}$, where it follows from the fact that $\pi_1(X \times Y) = \pi_1(X) \times \pi_1(Y)$.) By choosing a K-rational point c of C, on then gets a covering $W_c \longrightarrow V$ defined over K which is K'-isomorphic to W.

Step 3: Using th. 6.3.3, we may replace $(K, \{z_n\})$ by $(K_1, \{z_n\})$, where tr.deg$K_1/\mathbf{Q} < \infty$. One can then find an embedding $K_1 \longrightarrow \mathbf{C}$ which transforms the $\{z_n\}$ into the $\{e^{2\pi i/n})\}$ (irreducibility of the cyclotomic polynomials). Another application of th. 6.3.3 concludes the proof.

Remark: Let Γ be a discrete group, $\hat{\Gamma}$ its profinite completion

$$\hat{\Gamma} = \varprojlim \Gamma/N$$

where N runs over the normal subgroups of Γ of finite index. The canonical homomorphism $\Gamma \longrightarrow \hat{\Gamma}$ is universal for maps of Γ into profinite groups. *It is not always true that this map is injective.* There are examples of finitely presented Γ with Γ infinite but $\hat{\Gamma} = \{1\}$, e.g. the group defined by four generators x_i ($i \in \mathbf{Z}/4\mathbf{Z}$) with the four relations $x_i x_{i+1} x_i^{-1} = x_{i+1}^2$ (G. Higman).

There are examples of complex algebraic varieties with a fundamental group Γ such that the map $\Gamma \to \hat{\Gamma}$ is not injective, cf. [To].

In the case of curves, which is the case we are mostly interested in, the map $\pi_1(g, k) \longrightarrow \hat{\pi}_1(g, k)$ is injective. This follows from the fact that $\pi_1(g, k)$ is isomorphic to a subgroup of a linear group (e.g., $\mathbf{SL}_3(\mathbf{R})$ or even better, $\mathbf{SL}_2(\mathbf{R})$); one then uses a well-known result of Minkowski and Selberg (see e.g. [Bor], p. 119).

6.4   Appendix: universal ramified coverings of Riemann surfaces with signature

Let X be a Riemann surface, $S \subset X$ a finite set of points in X. Let us assign to each point $P \in S$ an integer n_P, with $2 \leq n_P < \infty$. Such a set S along with a set of integers n_P is called a *signature* on X.

One defines a ramified covering subordinate to the signature $(S; n_P)$ to be a holomorphic map $f : Y \longrightarrow X$, with the following properties:

1. If $S_Y = f^{-1}(S)$, the map $f : Y - S_Y \longrightarrow X - S$ is a topological covering.

2. Let $P \in S$ and choose a disk D in X with $D \cap S = \{P\}$. Then $f^{-1}(D)$ splits into connected components D_α, and the restriction f_α of f to D_α is isomorphic to the map $z \mapsto z^{n_\alpha}$, for some n_α dividing n_P.

(When $Y \longrightarrow X$ is a finite map, condition 2 just means that the ramification above P divides n_P.)

For every $P \in S$, let C_P be the conjugacy class "turning around P" in the fundamental group $\pi_1(X - S)$. If $Y \longrightarrow X$ is as above, $Y - f^{-1}(S) \longrightarrow X - S$ is a covering, and the action of $C_P^{n_P}$ is trivial. Conversely, given a covering of $X - S$ with that property, one shows (by an easy local argument) that it extends uniquely to a map $Y \longrightarrow X$ of the type above.

Theorem 6.4.1 *Let X, S, and n_P be given. Then there is a universal covering Y subordinate to the signature $(S; n_P)$ with faithful action of*

$$\Gamma = \pi_1(X - S)/N,$$

where N is the normal subgroup generated by the s^{n_P}, $s \in C_P$, $P \in S$. This Y is simply connected, and one has $Y/\Gamma = X$.

Sketch of proof: Let Y_N be the covering of $X - S$ associated to N. This is a Galois covering, with Galois group $\pi_1(X - S)/N$. As above, one can complete it to a covering $Y \longrightarrow X$. One checks that Y is universal, simply connected, and that Γ acts properly on Y, with $X = Y/\Gamma$.

We apply this to the case $X = \bar{X} - T$, where $\bar{X}$ is a compact Riemann surface of genus g, and T is a finite subset of $\bar{X}$. Letting $S = \{P_1, \ldots, P_s\}$ and $T = \{P_{s+1}, \ldots, P_{s+t}\}$, one associates to X a "generalized signature" on $\bar{X}$, where the indices are allowed to take to value ∞, by setting

$$\begin{cases} n_i = n_{P_i} & \text{if } i \le s, \\ n_i = \infty & \text{otherwise.} \end{cases}$$

The corresponding group Γ is then defined by generators $a_1, b_1, \ldots, a_g$, $b_g, c_1, \ldots c_{s+t}$ and relations:

$$\begin{cases} a_1 b_1 a_1^{-1} b_1^{-1} \cdots a_g b_g a_g^{-1} b_g^{-1} c_1 \cdots c_{s+t} = 1 \\ c_1^{n_1} = 1, \ldots, c_s^{n_s} = 1. \end{cases}$$

Example: In the case $g = 0$ and $s + t = 3$, the group Γ and the corresponding universal covering $\bar{X} \longrightarrow \mathbf{P}_1(\mathbf{C})$ can be constructed geometrically as follows. Set $\lambda = 1/n_1 + 1/n_2 + 1/n_3$ (with the convention that $1/\infty = 0$). If $\lambda > 1$, let $\bar{X}$ be the Riemann sphere $\mathbf{S}_2$ equipped with its metric of curvature 1; if $\lambda = 1$, let $\bar{X}$ be $\mathbf{C}$ with the euclidean metric; and if $\lambda < 1$, let $\bar{X}$ be the Poincaré upper half plane, equipped with the metric of curvature -1. In each case, by the Gauss-Bonnet theorem there is a geodesic triangle on $\bar{X}$ with angles π/n_1, π/n_2, and π/n_3. (The case $n_i = \infty$ gives an angle equal to 0, and the corresponding vertex is a "cusp" at infinity.) Denoting by s_i the reflection about the ith side ($i = 1, 2, 3$), the elements $C_1 = s_2 s_3$, $C_2 = s_3 s_1$, and $C_3 = s_1 s_2$ satisfy the relation

$$C_1 C_2 C_3 = 1$$

and the C_i are of order n_i. Hence the group Γ' generated by the C_i is a quotient of Γ, by the map sending c_i to C_i ($i = 1, 2, 3$). In fact, one can show that the projection $\Gamma \longrightarrow \Gamma'$ is an isomorphism, and that $\bar{X}$ with this action of Γ is the universal covering of th. 6.4.1. A fundamental domain for the action of Γ is given by the union of the geodesic triangle with one of its reflections about a side.

Theorem 6.4.2 *The element c_i of Γ has order n_i, except in the following two cases:*
  1. $g = 0$, $s + t = 1$;
  2. $g = 0$, $s + t = 2$, $n_1 \ne n_2$.

Consider first the case $g = 0$. This case can itself be divided into three subcases:

1. $s + t = 2$, and $n_1 = n_2$. In this case Γ is isomorphic to a cyclic group of order n_1 (if $n_1 = \infty$, then $\Gamma = \mathbf{Z}$), and the statement follows.

2. $s + t = 3$. The explicit construction of Γ above shows that the c_i are of order n_i.

3. $s + t \geq 4$. One reduces this to case 2 by adding the relations $c_i = 1$ for a set I of indices with $|I| = s + t - 3$. One thus gets a surjective homomorphism onto a group of the type treated in case 2, and the c_i with $i \notin I$ are therefore of order n_i. By varying I, one obtains the result.

To treat the case $g \geq 1$, it suffices to exhibit a group containing elements $A_1, \ldots, A_g, B_1, \ldots, B_g, C_1, \ldots, C_{s+t}$ satisfying the given relation for Γ, with the C_i of order n_i, $(i = 1, \ldots s+t)$. For this, one may choose elements C_i of order n_i in the special orthogonal group $\mathbf{SO}_2$. Every element x of $\mathbf{SO}_2$ can be written as a commutator (y, z) in the orthogonal group $\mathbf{O}_2$; for, choosing $y \in \mathbf{SO}_2$ such that $y^2 = x$, and $z \in \mathbf{O}_2 - \mathbf{SO}_2$, one has $yzy^{-1}z^{-1} = y \cdot y = x$. Hence, it suffices to take $A_j, B_j = 1$ if $j > 1$, and $A_1, B_1 \in \mathbf{O}_2$ such that

$$(A_1, B_1) = (C_1 C_2 \cdots C_{s+t})^{-1}.$$

Assume that the signature $(n_1, \ldots, n_{s+t})$ is not one of the exceptional cases in the theorem above. We then have a "universal" ramified covering Y which is simply connected. By the uniformization theorem (cf. [Fo]), such a Y is isomorphic to either $\mathbf{P}_1(\mathbf{C})$, $\mathbf{C}$, or the upper half plane $\mathcal{H}$. To tell which of these three cases occur, let us introduce the "Euler-Poincaré characteristic" E of the signature:

$$E = 2 - 2g - \sum_{i=1}^{s+t}(1 - \frac{1}{n_i}),$$

with the convention that $\frac{1}{\infty} = 0$.

Theorem 6.4.3

1. If $E > 0$, then Y is isomorphic to $\mathbf{P}_1(\mathbf{C})$.
2. If $E = 0$, then Y is isomorphic to $\mathbf{C}$.
3. If $E < 0$, then Y is isomorphic to the upper half plane $\mathcal{H}$.

This can be proved using the Gauss-Bonnet formula.

1. Case 1 occurs only if $g = 0$ and the signature is:

   - empty, corresponding to the identity covering $\mathbf{P}_1(\mathbf{C}) \longrightarrow \mathbf{P}_1(\mathbf{C})$.

   - $(n, n), n < \infty$, corresponding to the cyclic covering $z \mapsto z^n$.

- $(2, 2, n)$, corresponding to $\Gamma = D_n$, the dihedral group of order $2n$.

- $(2, 3, 3)$, corresponding to $\Gamma = A_4$ (tetrahedral group).

- $(2, 3, 4)$, corresponding to $\Gamma = S_4$ (octahedral group).

- $(2, 3, 5)$, corresponding to $\Gamma = A_5$ (icosahedral group).

2. Case 2 occurs for

- $g = 1$ and empty signature, corresponding to the covering of a complex torus by the complex plane, $\Gamma = $ a lattice.

- $g = 0$ and signature (∞, ∞), corresponding to the map

$$\exp : \mathbf{C} \longrightarrow \mathbf{C}, \quad \Gamma = \mathbf{Z}.$$

- signature $(2, 4, 4)$, corresponding to $\Gamma = $ group of affine motions preserving the lattice $\mathbf{Z} + \mathbf{Z}i$.

- signature $(2, 3, 6)$, corresponding to $\Gamma = $ group of affine motions preserving the lattice $\mathbf{Z} + (\frac{1+\sqrt{-3}}{2})\mathbf{Z}$.

- signature $(3, 3, 3)$, corresponding to a subgroup of index 2 of the previous one.

- signature $(2, 2, 2, 2)$, corresponding to coverings

$$\mathbf{C} \longrightarrow T \longrightarrow \mathbf{P}_1,$$

where T is an elliptic curve and $\mathbf{C} \longrightarrow T$ is the universal unramified covering map.

3. Case 3 corresponds to the case where Γ is a discrete subgroup of $\mathbf{PSL}_2(\mathbf{R})$ which is a "Fuchsian group of the first kind", cf. e.g. Shimura [Shim], Chap. 1; one then has $X = \mathcal{H}/\Gamma$ and $\bar{X}$ is obtained by adding to X the set of cusps (mod Γ). (By a theorem of Siegel [Sie], such groups Γ can be characterized as the discrete subgroups of $\mathbf{PSL}_2(\mathbf{R})$ with finite covolume.) The generators c_i with $n_i < \infty$ are elliptic elements of the Fuchsian group Γ, the c_i with $n_i = \infty$ correspond to parabolic elements, and the a_j and b_j to hyperbolic elements.

Chapter 7

Rigidity and rationality on finite groups

7.1 Rationality

Let G be a finite group, $\mathrm{Cl}(G)$ the set of its conjugacy classes. Choose N such that every element of G has order dividing N; the group $\Gamma_N = (\mathbf{Z}/N\mathbf{Z})^*$ acts on G by sending s to s^α, for $\alpha \in (\mathbf{Z}/N\mathbf{Z})^*$, and acts similarly on $\mathrm{Cl}(G)$.

Let $X(G)$ be the set of irreducible characters of G. They take values in $\mathbf{Q}(\mu_N)$, and hence there is a natural action of $\mathrm{Gal}(\mathbf{Q}(\mu_N)) \simeq \Gamma_N$ on $X(G)$. The actions of Γ_N on $\mathrm{Cl}(G)$ and on $X(G)$ are related by the formula

$$\sigma_\alpha(\chi)(s) = \chi(s^\alpha),$$

where $\sigma_\alpha \in \mathrm{Gal}(\mathbf{Q}(\mu_N)/\mathbf{Q})$ is the element sending the Nth roots of unity to their αth powers.

Definition 7.1.1 *A class c in $\mathrm{Cl}(G)$ is called $\mathbf{Q}$-rational if the following equivalent properties hold:*
1. c is fixed under Γ_N.
2. Every $\chi \in X(G)$ takes values in $\mathbf{Q}$ (and hence in $\mathbf{Z}$) on c.
(The equivalence of these two conditions follows from the formula above.)

The rationality condition means that, if $s \in c$, then all of the generators of the cyclic group $<s>$ generated by s are in c, i.e., are conjugate to s.

For instance, in the symmetric group S_n, every conjugacy class is rational.

More generally, let K be any field of characteristic zero. The class c of an element $s \in G$ is called K-rational if $\chi(s) \in K$ for all $\chi \in X(G)$, or equivalently, if $c^\alpha = c$ for all α such that $\sigma_\alpha \in \mathrm{Gal}(K(\mu_N)/K)$.

For example, the alternating group A_5 has five conjugacy classes of order 1, 2, 3, 5, and 5 respectively. Let us call $5A$ and $5B$ the two conjugacy classes of exponent 5. (They can be distinguished as follows: given a five-cycle s, let $\alpha(s)$ be the permutation of $\{1, 2, 3, 4, 5\}$ sending i to $s^i(1)$. Then s is in the conjugacy class $5A$ if and only if $\alpha(s)$ is even.) If $s \in 5A$, then $s^{-1} \in 5A$ and $s^2, s^3 \in 5B$, so the classes $5A$ and $5B$ are not rational over $\mathbf{Q}$. However, they are rational over the quadratic field $\mathbf{Q}(\sqrt{5})$.

Remark: The formula $\sigma_\alpha(\chi)(s) = \chi(s^\alpha)$ has an analogue for supercuspidal admissible representations of semi-simple p-adic groups. This can be deduced from the fact, proved by Deligne (cf. [De1]), that the character of such a representation is supported by the elliptic elements (note that these are the only elements s for which the notation s^α makes sense).

Rationality of inertia

Let K be a local field with residue field k of characteristic zero, and let M be a Galois extension of K with group G. There is a unique maximal unramified extension L of K in M; it is Galois over K. Let I be the inertia subgroup of G, i.e., $I = \mathrm{Gal}(M/L)$. Since the ramification is tame, I is cyclic. In fact, there is a canonical isomorphism

$$I \xrightarrow{\sim} \mu_e(l),$$

where e denotes the ramification index of L/K, and l is the residue field of L. (This isomorphism sends $\sigma \in I$ to $\sigma\pi/\pi \pmod{\pi}$, where π is a uniformizing element of M.)

Proposition 7.1.2 *The class in G of an element of I is rational over k.*

Indeed, the group $H = \mathrm{Gal}(L/K) = \mathrm{Gal}(l/k)$ acts on $\mu_e(l)$ in a natural way, and on I by conjugation. These actions are compatible with the isomorphism above. If now σ_α is an element of $\mathrm{Gal}(k(\mu_e)/k)$, there exists $t \in H$ such that t acts on $\mu_e(l)$ by $z \mapsto z^\alpha$. If $s \in I$, then s^α and tst^{-1} have the same image in $\mu_e(l)$. Hence $s^\alpha = tst^{-1}$. This shows that the class c of s is such that $c^\alpha = c$, q.e.d.

Corollary 7.1.3 *If $k = \mathbf{Q}$, then the classes in I are rational in G.*

Remark on the action of Γ_N on $\mathrm{Cl}(G)$ and $X(G)$

The Γ_N-sets $\mathrm{Cl}(G)$ and $X(G)$ are the character sets of the étale $\mathbf{Q}$-algebras $\mathbf{Q} \otimes R(G)$ and $Z\mathbf{Q}[G]$, where $R(G)$ is the representation ring of

G (over $\bar{\mathbf{Q}}$), and $Z\mathbf{Q}[G]$ is the center of the group algebra $\mathbf{Q}[G]$. These Γ_N-sets are easily proved to be "weakly isomorphic" in the sense of exercise 1 below. However, they are not always isomorphic (see [Th1]).

Exercises:
1. Let X and Y be finite sets on which acts a finite group Γ.
   (a) Show the equivalence of the following properties:
   i. $|X^C| = |Y^C|$ for every cyclic subgroup C of Γ;
   ii. $|X/C| = |Y/C|$ for every cyclic subgroup C of Γ;
   iii. $|X/H| = |Y/H|$ for every subgroup H of Γ;
   iv. The $\mathbf{Q}$-linear representations of Γ defined by X and Y are isomorphic.
If those properties are true, the Γ-sets X and Y are said to be weakly isomorphic.
   (b) Show that weak isomorphism is equivalent to isomorphism when Γ is cyclic ("Brauer's Lemma"). Give an example where Γ is a non-cyclic group of order 4, $|X| = |Y| = 6$, and X and Y are weakly isomorphic but not isomorphic.

2. Use exerc. 1.(b) to prove that, if G is a p-group, $p \neq 2$, then $\mathbf{Q} \otimes R(G)$ is isomorphic to $Z\mathbf{Q}[G]$.

3. Extend prop. 7.1.2 to the case where k has characteristic $p > 0$, assuming that M/K is tame.

7.2 Counting solutions of equations in finite groups

Let G be a compact group, equipped with its Haar measure of total mass 1. Let χ be an irreducible character of G, with $\rho : G \longrightarrow \mathbf{GL}(E)$ the corresponding linear representation. By Schur's lemma, $\int_G \rho(txt^{-1})\,dt$ is a scalar multiple $\lambda \cdot 1_E$ of the identity in $\mathbf{GL}(E)$. Taking traces gives $\chi(x) = \lambda\chi(1)$, hence

$$\int_G \rho(txt^{-1})dt = \frac{\chi(x)}{\chi(1)}1_E.$$

Multiplying on the right by $\rho(y)$, we get:

$$\int_G \rho(txt^{-1}y)dt = \frac{\chi(x)}{\chi(1)}\rho(y).$$

Taking traces gives:

$$\int_G \chi(txt^{-1}y)dt = \frac{\chi(x)\chi(y)}{\chi(1)}.$$

This formula extends by induction to k elements $x_1, \ldots, x_k$:

$$\int_G \cdots \int_G \chi(t_1 x_1 t_1^{-1} \cdots t_k x_k t_k^{-1} y) \, dt_1 \, dt_2 \cdots dt_k = \frac{\chi(x_1) \cdots \chi(x_k) \chi(y)}{\chi(1)^k}$$
$$(7.1)$$

More generally, let ϕ be a class function on G,

$$\phi = \sum_\chi c_\chi \chi,$$

where c_χ is the inner product of ϕ with the irreducible character χ,

$$c_\chi = \int_G \phi(x) \bar{\chi}(x) \, dx,$$

and assume that $\sum |c_\chi| \chi(1) < \infty$ so that the sum is normally convergent. By summing over all characters in formula 7.1, we see that the integral

$$I(\phi) = \int_G \cdots \int_G \phi(t_1 x_1 t_1^{-1} \cdots t_k x_k t_k^{-1} y) dt_1 \, dt_2 \cdots dt_k$$

is equal to

$$\sum_\chi c_\chi \frac{\chi(x_1) \cdots \chi(x_k) \chi(y)}{\chi(1)^k}.$$

Let us compute $I(\phi)$ in the case where G is a finite group and ϕ is the Dirac function which is 1 at the identity element, and 0 elsewhere. One has

$$\phi = \frac{1}{|G|} \sum_\chi \chi(1) \chi,$$

and hence $c_\chi = \chi(1)/|G|$. Given elements $x_1, \ldots, x_k$, and y in G, the value of $I(\phi)$ is $N/|G|^k$, where N is the number of solutions $(t_1, \ldots, t_k)$ of the equation

$$t_1 x_1 t_1^{-1} \cdots t_k x_k t_k^{-1} y = 1.$$

Hence

$$N = |G|^k \sum_\chi \frac{\chi(1)}{|G|} \frac{\chi(x_1) \cdots \chi(x_k) \chi(y)}{\chi(1)^k}$$

$$= |G|^{k-1} \sum_\chi \frac{\chi(x_1) \cdots \chi(x_k) \chi(y)}{\chi(1)^{k-1}}.$$
$$(7.2)$$

Let $C_1, \ldots, C_k$ denote the conjugacy classes of the elements $x_1, \ldots, x_k$, and define $n(C_1, \ldots, C_k)$ to be the number of solutions $(g_1, \ldots, g_k)$ of the equation

$$g_1 g_2 \cdots g_k = 1, \quad g_i \in C_i.$$

Letting Z_i be the order of the centralizer of an element of C_i, one has

$$n(C_1, \ldots, C_k) = \frac{N}{Z_1 \cdots Z_k}.$$

By applying formula 7.2 and observing that $Z_i = |G|/|C_i|$, one therefore gets:

Theorem 7.2.1 *The number $n = n(C_1, \ldots, C_k)$ of solutions of the equation $g_1 \cdots g_k = 1$, $g_i \in C_i$, is given by*

$$n = \frac{1}{|G|} |C_1| \cdots |C_k| \sum_{\chi} \frac{\chi(x_1) \cdots \chi(x_k)}{\chi(1)^{k-2}},$$

where x_i is a representative of the conjugacy class C_i, and χ runs through all the irreducible characters of G.

Exercises:

1. Let G be a compact group and let ρ be an irreducible representation of G with character χ. Show that

$$\int \int \rho(txt^{-1}x^{-1}) dx\, dt = 1/\chi(1)^2.$$

(Hint: Show that the left hand side is an intertwining operator for ρ, and hence is a scalar by Schur's lemma. Then compute its trace.) Conclude that

$$\int \int \rho(txt^{-1}x^{-1} \cdot y) dx\, dt = \rho(y)/\chi(1)^2, \qquad \text{for all } y \in G,$$

and that

$$\int \int \chi(txt^{-1}x^{-1} \cdot y) dx\, dt = \chi(y)/\chi(1)^2.$$

Hence show by induction that:

$$\int \cdots \int \chi(t_1 x_1 t_1^{-1} x_1^{-1} \cdots t_g x_g t_g^{-1} x_g^{-1} \cdot y) dx_1\, dt_1 \cdots dx_g\, dt_g = \chi(y)/\chi(1)^{2g}.$$

2. Suppose now that G is finite. We denote the commutator $uvu^{-1}v^{-1}$ of u and v by (u, v). Let $g \geq 0$, and let z_j be fixed elements of G, for $j = 1, \ldots, k$. For $y \in G$, let $N = N(g, z_j, y)$ denote the number of tuples of elements of G,

$$(u_1, v_1, \ldots, u_g, v_g, t_1, \ldots, t_k),$$

such that

$$(u_1, v_1) \cdots (u_g, v_g) t_1 z_1 t_1^{-1} \cdots t_k z_k t_k^{-1} = y^{-1}.$$

Using exerc. 1, show that:

$$N = |G|^{2g+k-1} \sum_{\chi} \chi(z_1) \cdots \chi(z_k) \chi(y)/\chi(1)^{2g+k-1}.$$

3. Show that an element y of G is a product of g commutators if and only if the sum

$$\sum_\chi \chi(y)/\chi(1)^{2g-1}$$

is non-zero. In particular, y is a commutator if and only if $\sum \chi(y)/\chi(1) \neq 0$. It is a known conjecture of Öystein Ore that every element in a finite non-abelian simple group is a commutator. The reader may wish to verify this (for one of the sporadic groups, say) by using the formula above and the character tables in [ATLAS].

Remark: Th. 7.2.1 can be used to compute the number of subgroups in a finite group which are isomorphic to the alternating group A_5. Indeed, A_5 has a presentation given by three generators x, y, z, and relations

$$x^2 = y^3 = z^5 = xyz = 1.$$

The problem therefore amounts to finding the number of solutions of the equation $xyz = 1$, where x, y z belong to conjugacy classes of exponent 2, 3, and 5 respectively. The same remark applies to S_4, A_4, and the dihedral groups, which have similar presentations, cf. th. 6.4.3, case 1.

7.3   Rigidity of a family of conjugacy classes

Let G be a finite group, and fix conjugacy classes C_1, ..., C_k in G. Let $\bar{\Sigma} = \bar{\Sigma}(C_1, \ldots, C_k)$ be the set of all $(g_1, \ldots, g_k)$, with $g_i \in C_i$, such that $g_1 \cdots g_k = 1$. (Hence, by the previous section, $n(C_1, \ldots, C_k) = |\bar{\Sigma}|$.) Let $\Sigma = \Sigma(C_1, \ldots, C_k)$ be the set of $(g_1, \ldots, g_k)$ in $\bar{\Sigma}$ such that g_1, ..., g_k generate the group G. The group G acts by conjugation on Σ and $\bar{\Sigma}$.

Assume that the center of G is trivial (as is the case, for example, when $G = S_n$, $n \geq 3$, or when G is a non-abelian simple group). Then *the action of G on Σ is free*: for, if $g \in G$ fixes $(g_1, \ldots, g_k)$, it commutes with the g_i's, and hence with all of G, since the g_i's generate G; but then $g = 1$, because G is assumed to have trivial center.

One says that a k-tuple of conjugacy classes $(C_1, \ldots, C_k)$ is *rigid* if $\Sigma \neq \emptyset$ and G acts transitively on Σ, i.e.,

$$|\Sigma| = |G|.$$

One says that $(C_1, \ldots, C_k)$ is *strictly rigid* if it is rigid, and $\Sigma = \bar{\Sigma}$.

The order of $\bar{\Sigma}$ can be computed using the formula of the previous section:

$$|\bar{\Sigma}| = \frac{1}{|G|}|C_1| \cdots |C_k| \sum_\chi \chi(c_1)\chi(c_2) \cdots \chi(c_k)/\chi(1)^{k-2},$$

where $c_i \in C_i$ for $i = 1, \ldots, k$, and χ runs through the irreducible characters of G. If z_i is the order of the centralizer $Z(c_i)$, this can also be written as:

$$|\bar{\Sigma}| = \frac{|G|^{k-1}}{z_1 \ldots z_k} \sum_{\chi} \chi(c_1)\chi(c_2)\cdots\chi(c_k)/\chi(1)^{k-2}. \tag{7.3}$$

Rigidity is often proved in two steps:
1. Compute the order of $\bar{\Sigma}$, by using formula 7.3 and the character table of G.
2. Compute the order of $\bar{\Sigma} - \Sigma$, by finding k-tuples in $\bar{\Sigma}$ which do not generate all of G: for this, one uses a knowledge of the maximal subgroups of G (whenever possible).

Remarks:
1. Let σ be an outer automorphism of G. Suppose $(C_1, \ldots, C_k)$ is a rigid k-tuple of conjugacy classes. Then $\sigma(C_i) \neq C_i$ for some i. Indeed, suppose that σ preserves each C_i. Then, letting $(g_1, \ldots, g_k)$ be an element of Σ, it follows that $(\sigma g_1, \ldots, \sigma g_k)$ belongs to Σ. Since G acts transitively on Σ by inner conjugation, there exists $g \in G$ such that

$$\sigma g_i = g g_i g^{-1} \qquad \text{for all } i.$$

But the g_i generate G and hence σ is an inner automorphism, which contradicts the assumption.

2. In many cases, the term $\sum_{\chi} \chi(C_1)\cdots\chi(C_k)/\chi(1)^{k-2}$ in formula 7.3 is not very large, the main contribution to the sum being given by the unit character $\chi = 1$. For a rigid family of conjugacy classes, one might therefore expect that the order of magnitude of $|C_1|\cdots|C_k|$ is close to $|G|$.

As in §7.1, let $\alpha \in (\mathbf{Z}/N\mathbf{Z})^*$, where N is a multiple of the exponent of G, so that $(\mathbf{Z}/N\mathbf{Z})^*$ acts on $\mathrm{Cl}(G)$. Then one has:
Proposition 7.3.1
1. $|\bar{\Sigma}(C_1^\alpha, \ldots, C_k^\alpha)| = |\bar{\Sigma}(C_1, \ldots, C_k)|$;
2. $|\Sigma(C_1^\alpha, \ldots, C_k^\alpha)| = |\Sigma(C_1, \ldots, C_k)|$.

The first identity follows from formula 7.3 above, combined with the formula $\chi(c^\alpha) = \sigma_\alpha(\chi)(c)$ of §7.1. The second is proved by induction on the order of G, as follows: if H is a subgroup of G, let $\Sigma^{(H)}(C_1 \cap H, \ldots, C_k \cap H)$ denote the set of $(g_1, \ldots, g_k)$, with $g_i \in C_i \cap H$ for all i, such that $g_1 \cdots g_k = 1$ and the g_i generate H. (In general, the $C_i \cap H$ are not conjugacy classes in H, but unions of conjugacy classes.) The formula

$$\bar{\Sigma} - \Sigma(C_1, \ldots, C_k) = \bigcup_{H \subset G, \ H \neq G} \Sigma^{(H)}(C_1 \cap H, \ldots, C_k \cap H)$$

supplies the induction step.

Corollary 7.3.2 *If* $(C_1, \ldots, C_k)$ *is rigid (resp. strictly rigid), so is the family* $(C_1^\alpha, \ldots, C_k^\alpha)$.

Remark: Here is another way to prove prop. 7.3.1. Let F be the group with presentation given by generators $x_1, \ldots, x_k$ and relation $x_1 \cdots x_k = 1$, and let $\hat{F}$ be its profinite completion. Then one has:

Proposition 7.3.3 *For each* $\alpha \in \hat{\mathbf{Z}}^*$, *there is an automorphism* θ *of* $\hat{F}$ *such that* $\theta(x_i)$ *belongs to the conjugacy class of* x_i^α. (Equivalently, there exist elements $y_1, \ldots, y_k$ with y_i conjugate to x_i^α, satisfying $y_1 \cdots y_k = 1$, and generating $\hat{F}$.)

Let us use the interpretation of $\hat{F}$ as the algebraic fundamental group π of the projective line $\mathbf{P}_1$ with k points removed. By choosing a coherent system $\{z_n\}$ of roots of unity in $\bar{\mathbf{Q}}$, one has an identification (cf. §6.3)

$$\hat{F} \overset{\sim}{\longrightarrow} \pi.$$

Replacing the system $\{z_n\}$ by $\{z_n^\alpha\}$ gives a different isomorphism, and composing the two yields the desired automorphism θ of $\hat{F}$.

Prop. 7.3.3 implies that $|\Sigma(C_1^\alpha, \ldots, C_k^\alpha)| = |\Sigma(C_1, \ldots, C_k)|$. For, the elements of $\Sigma(C_1, \ldots, C_k)$ are in one-one correspondence with the surjective homomorphisms $\hat{F} \longrightarrow G$ sending each x_i to an element of C_i, while the elements of $\Sigma(C_1^\alpha, \ldots, C_k^\alpha)$ correspond to the surjective homomorphisms sending each x_i to an element in C_i^α. The automorphism $\theta : \hat{F} \longrightarrow \hat{F}$ induces a map $\mathrm{Hom}(\hat{F}, G) \longrightarrow \mathrm{Hom}(\hat{F}, G)$ which gives a bijection between the two sets $\Sigma(C_1, \ldots, C_k)$ and $\Sigma(C_1^\alpha, \ldots, C_k^\alpha)$.

Exercise: Show that, if $(C_1, \ldots, C_k)$ is rigid (resp. strictly rigid), then so is the family $(C_{\sigma(1)}, \ldots, C_{\sigma(k)})$ for any permutation σ of $\{1, \ldots, k\}$.

7.4   Examples of rigidity

We give only a few such examples. For more, the reader should consult [Ma3] and [MM].

7.4.1   The symmetric group S_n

The symmetric group S_n ($n \geq 3$) has conjugacy classes nA, $2A$, and $C^{(1)}$ corresponding to cycles of order n, 2, and $n - 1$ respectively. *The triple* $(nA, 2A, C^{(1)})$ *is strictly rigid.* For, giving an n-cycle $x \in nA$ determines a cyclic arrangement of $\{1, \ldots, n\}$ (i.e., an oriented n-gon). Composing this permutation of order n with a transposition gives an $(n-1)$-cycle if and only if the two vertices which are permuted are consecutive. Hence, the solutions x, y, z of the equation $xyz = 1$, with x, y, z cycles of order n, 2 and $n-1$ respectively, are in one to one correspondence with the oriented n-gons with a distinguished edge. Any two such configurations can be tranformed into one another by a unique permutation in S_n; hence $|\bar{\Sigma}| = |G|$. But $\bar{\Sigma} = \Sigma$, since (12) and $(12\ldots n)$ are known to generate S_n; this shows that $(nA, 2A, C^{(1)})$ is strictly rigid.

More generally, consider the conjugacy classes nA, $2A$, and $C^{(k)}$, where $C^{(k)}$ is the class of the permutation

$$(1\ldots k)(k+1\ldots n).$$

As before, an element $(x, y, z) \in \bar{\Sigma}$ corresponds to an oriented n-gon with two distinguished vertices separated by k edges. If $k \neq n/2$, any two such configurations can be transformed into one another by a unique permutation in S_n, and hence $|\bar{\Sigma}| = |G|$. However, $(nA, 2A, C^{(k)})$ is not rigid in general; to get rigidity, one must assume that $(k, n) = 1$. In that case, any triple (x, y, z) in $\bar{\Sigma}$ generates S_n: by relabelling if necessary, we may write $x = (1 \ldots n)$, and $y = (1, k+1)$. Since $(k, n) = 1$, the permutation x^k is still an n-cycle. By relabelling again, the group generated by x^k and y is isomorphic to the group generated by the permutations $(1 \ldots n)$ and (12). This in turn contains the group generated by (12), (23), ..., $(n-1, n)$, which is equal to S_n by a well-known result (cf. §4.4).

Exercise: Check that $|\bar{\Sigma}| = n!$ for the conjugacy classes $(2A, nA, C^{(1)})$ by applying formula 7.3 of §7.3 (prove that the only non-zero terms come from the two characters of degree 1 of S_n).

7.4.2   The alternating group A_5

The alternating group A_5 has unique conjugacy classes of order 2 and 3, denoted by $2A$ and $3A$ respectively. It has two conjugacy classes of order 5 which are rational over $\mathbf{Q}(\sqrt{5})$ and conjugate to each other, denoted by $5A$ and $5B$.

Proposition 7.4.1 *The following triples of conjugacy classes are strictly rigid:* $(2A, 3A, 5A), (2A, 5A, 5B),$ *and* $(3A, 5A, 5B)$.

To prove this, one can compute the order of $\bar{\Sigma}$ in each case from the character table of A_5. In ATLAS style, it is:

characters ↓	60 1A	4 2A	3 3A	5 5A	5 5B	orders of ←centralizers ← classes
χ_1	1	1	1	1	1	
χ_2	3	−1	0	z'	z	$z = \frac{1+\sqrt{5}}{2}$
χ_3	3	−1	0	z	z'	$z' = \frac{1-\sqrt{5}}{2}$
χ_4	4	0	1	−1	−1	
χ_5	5	1	−1	0	0	

One then gets:

$$|\bar{\Sigma}(2A, 3A, 5A)| = \frac{60^2}{4 \cdot 3 \cdot 5}(1 + 0 + 0 + 0 + 0) = 60$$

$$|\bar{\Sigma}(2A, 5A, 5B)| = \frac{60^2}{4 \cdot 5 \cdot 5}(1 + \frac{1}{3} + \frac{1}{3} + 0 + 0) = 60$$

$$|\bar{\Sigma}(3A, 5A, 5B)| = \frac{60^2}{3 \cdot 5 \cdot 5}(1 + 0 + 0 + \frac{1}{4} + 0) = 60.$$

One checks easily that any triple in any of these $\bar{\Sigma}$ generates A_5, and prop. 7.4.1 follows.

Exercises:
1. Show that the triples $(3A, 3A, 5A)$, $(3A, 5A, 5A)$, and $(5A, 5A, 5A)$ are strictly rigid.
2. Show that $(2A, 2A, 5A)$ is not rigid, even though $|\bar{\Sigma}| = 60$ in that case (the triples in $\bar{\Sigma}$ generate dihedral subgroups of order 10).

7.4.3   The groups $\mathbf{PSL_2(F_p)}$

The group $\mathbf{PSL_2(F_p)}$, with $p > 2$, contains unique conjugacy classes of elements of order 2 and 3, denoted by $2A$ and $3A$ respectively. There are two classes pA and pB of elements of order p, which are represented by unipotent matrices, $\begin{pmatrix} 1 & 1 \\ 0 & 1 \end{pmatrix}$, and $\begin{pmatrix} 1 & \alpha \\ 0 & 1 \end{pmatrix}$, where $\left(\frac{\alpha}{p}\right) = -1$.

Proposition 7.4.2 *The triple* $(2A, 3A, pA)$ *is strictly rigid.*

One checks that $(x_0, y_0, z_0) \in \Sigma$, where x_0, y_0, and z_0 are represented by the matrices:

$$x_0 = \begin{pmatrix} 0 & 1 \\ -1 & 0 \end{pmatrix}, \qquad y_0 = \begin{pmatrix} 0 & -1 \\ 1 & -1 \end{pmatrix}, \qquad z_0 = \begin{pmatrix} 1 & 1 \\ 0 & 1 \end{pmatrix}$$

Conversely, let (x, y, z) be in $\bar{\Sigma}$. We lift x, y, z to $\tilde{x}, \tilde{y}, \tilde{z}$ in $SL_2(\mathbf{F}_p)$, with $\tilde{x}$ of order 4, $\tilde{y}$ of order 3 and $\tilde{z}$ of order p, so that we have $\tilde{x}\tilde{y}\tilde{z} = \pm 1$. We view these elements as automorphisms of the vector space $V = \mathbf{F}_p \oplus \mathbf{F}_p$. Let D be the line of V fixed by $\tilde{z}$ and let $D' = \tilde{x}D$ be its transform by $\tilde{x}$. One has $D' \neq D$ (otherwise, $\pm \tilde{x}\tilde{z}$ would not be of order 3). After conjugating by an element of $\mathbf{SL}_2(\mathbf{F}_p)$, we may assume that D (resp. D') is the first (resp. the second) axis of coordinates in V. This means that we have

$$\tilde{x} = \begin{pmatrix} 0 & -\lambda \\ \lambda^{-1} & 0 \end{pmatrix}, \quad \text{and } \tilde{z} = \begin{pmatrix} 1 & \mu \\ 0 & 1 \end{pmatrix},$$

for some λ, μ in $\mathbf{F}_p^*$. By assumption, z belongs to the class pA of $\begin{pmatrix} 1 & 1 \\ 0 & 1 \end{pmatrix}$; this implies that μ is a square. If we write $\mu = \nu^2$, and conjugate by $\begin{pmatrix} \nu & 0 \\ 0 & \nu^{-1} \end{pmatrix}$, we see that we can further assume that $\mu = 1$, i.e., that $\tilde{z} = z_0$. Moreover, since $\tilde{z}\tilde{x}$ is of order 3 or 6, we have $\text{Tr}\,(\tilde{z}\tilde{x}) = \pm 1$. This gives $\lambda = \pm 1$, hence $x = x_0$, $y = y_0$, and $z = z_0$, which proves the result.

Proposition 7.4.3 *The triple* $(2A, pA, pB)$ *is strictly rigid if* $\left(\frac{2}{p}\right) = -1$.

One checks that (x_0, y_0, z_0) is in Σ, where

$$x_0 = \begin{pmatrix} 1 & -1 \\ 2 & -1 \end{pmatrix}, \quad y_0 = \begin{pmatrix} 1 & 1 \\ 0 & 1 \end{pmatrix}, \quad z_0 = \begin{pmatrix} 1 & 0 \\ -2 & 1 \end{pmatrix}.$$

The element z_0 is conjugate to $\begin{pmatrix} 1 & 2 \\ 0 & 1 \end{pmatrix}$; since $\left(\frac{2}{p}\right) = -1$, it belongs to the class pB. Conversely, let $(x, y, z) \in \Sigma$. We lift (x, y, z) as above to $(\tilde{x}, \tilde{y}, \tilde{z})$ with $\tilde{x}$ of order 4 and $\tilde{y}, \tilde{z}$ of order p. Let D be the line fixed by $\tilde{y}$ and D' be the line fixed by $\tilde{z}$. We may again assume that these lines are the standard coordinate lines, and that

$$\tilde{y} = \begin{pmatrix} 1 & 1 \\ 0 & 1 \end{pmatrix}, \quad \tilde{z} = \begin{pmatrix} 1 & 0 \\ \lambda & 1 \end{pmatrix}.$$

Writing that $\tilde{y}\tilde{z}$ has order 4, one gets $\text{Tr}\,(\tilde{y}\tilde{z}) = 0$, i.e. $\lambda = -2$, q.e.d.

Proposition 7.4.4 *The triple* $(3A, pA, pB)$ *is strictly rigid if* $\left(\frac{3}{p}\right) = -1$.

One checks that (x_0, y_0, z_0) is in Σ, where

$$x_0 = \begin{pmatrix} 1 & -1 \\ 3 & -2 \end{pmatrix}, \quad y_0 = \begin{pmatrix} 1 & 1 \\ 0 & 1 \end{pmatrix}, \quad z_0 = \begin{pmatrix} 1 & 0 \\ -3 & 1 \end{pmatrix}.$$

The assumption that 3 is not a quadratic residue mod p ensures that z_0 is in the class pB.

Conversely, let (x, y, z) be in $\bar{\Sigma}$. Using liftings $(\tilde{x}, \tilde{y}, \tilde{z})$ as above, one may assume that

$$\tilde{y} = \begin{pmatrix} 1 & 1 \\ 0 & 1 \end{pmatrix}, \quad \tilde{z} = \begin{pmatrix} 1 & 0 \\ \lambda & 1 \end{pmatrix}.$$

Writing that $\tilde{y}\tilde{z}$ has order 3 or 6, one gets $\mathrm{Tr}\,(\tilde{y}\tilde{z}) = \pm 1$, i.e., $\lambda = -1$ or $\lambda = -3$. However, $\lambda = -1$ is impossible (it would imply that $\tilde{z}$ belongs to the class pA); hence $\lambda = -3$, q.e.d.

7.4.4　The group $\mathbf{SL}_2(\mathbf{F}_8)$

The simple group $G = \mathbf{SL}_2(\mathbf{F}_8)$, of order 504, has three distinct conjugacy classes of order 9, denoted $9A$, $9B$, and $9C$ which are rational over the cubic field $\mathbf{Q}(\cos\frac{2\pi}{9})$ and conjugate to each other (cf. [ATLAS], p.6).

Proposition 7.4.5 *The triple $(9A, 9B, 9C)$ is strictly rigid.*

The character table for G is ([ATLAS], *loc. cit.*):

	504	8	9	7	7	7	9	9	9	orders of ←centralizers
	$1A$	$2A$	$3A$	$7A$	$7B$	$7C$	$9A$	$9B$	$9C$	← classes
χ_1	1	1	1	1	1	1	1	1	1
χ_2	7	-1	-2	0	0	0	1	1	1
χ_3	7	-1	1	0	0	0	x	x'	x''
χ_4	7	-1	1	0	0	0	x''	x	x'
χ_5	7	-1	1	0	0	0	x'	x''	x
χ_6	8	0	-1	1	1	1	-1	-1	-1
χ_7	9	1	0	y	y'	y''	0	0	0
χ_8	9	1	0	y''	y	y'	0	0	0
χ_9	9	1	0	y'	y''	y	0	0	0

$$x = -2\cos\tfrac{2\pi}{9}, \qquad x' = -2\cos\tfrac{4\pi}{9}, \qquad x'' = -2\cos\tfrac{8\pi}{9}, \qquad xx'x'' = 1;$$
$$y = 2\cos\tfrac{2\pi}{7}, \qquad y' = 2\cos\tfrac{4\pi}{7}, \qquad y'' = 2\cos\tfrac{8\pi}{7}, \qquad yy'y'' = 1.$$

Using formula (7.3) of §7.3, one gets:

$$|\bar{\Sigma}(9A, 9B, 9C)| = \frac{504^2}{9^3}\left(1 + \frac{1}{7} + \frac{1}{7} + \frac{1}{7} + \frac{1}{7} - \frac{1}{8} + 0 + 0 + 0\right) = 504 = |G|.$$

Hence it suffices to show that any $(x, y, z) \in \bar{\Sigma}$ generates G to prove rigidity.

The only maximal subgroups of G containing an element of order 9 are the normalizers of the non-split Cartan subgroups, which are isomorphic to a semi-direct product $C_2(\mathbf{F}_{64}^*)_1$, where $(\mathbf{F}_{64}^*)_1$ denotes the multiplicative group of elements of $\mathbf{F}_{64}$ of norm 1 over $\mathbf{F}_8$, and the non-trivial element of C_2 acts on $(\mathbf{F}_{64}^*)_1$ by $x \mapsto x^{-1}$; they are dihedral groups of order 18.

If $(x, y, z) \in \bar{\Sigma}$ does not generate G, then x, y, z are contained in such a normalizer. It follows that (by interchanging y and z if necessary):

$$y = x^{\pm 2}, \qquad z = x^{\pm 4}.$$

But then $xyz = x^{1 \pm 2 \pm 4}$ is not equal to 1. This contradiction completes the proof.

Exercises:

1. Show that the triples $(7A, 7A, 7A)$, $(2A, 3A, 7A)$, $(2A, 3A, 9A)$ are strictly rigid, and that $(7A, 7B, 7C)$ is not rigid.

2. Let $G \cdot 3$ be the automorphism group of $\mathbf{SL}_2(\mathbf{F}_8)$, cf. [ATLAS], p. 6. Show that the triple $(9A, 3B, 3C)$ is strictly rigid; the class $9A$ is rational (as a class of $G \cdot 3$); the classes $3B$ and $3C$ are rational over $\mathbf{Q}(\sqrt{-3})$, and conjugate to each other.

7.4.5 The Janko group J_1

The sporadic simple group J_1 discovered by Janko is of order

$$175560 = 2^3 \cdot 3 \cdot 5 \cdot 7 \cdot 11 \cdot 19.$$

It contains conjugacy classes $2A$, $5A$, and $5B$ of orders 2, 5, and 5; the classes $5A$ and $5B$ are rational over $\mathbf{Q}(\sqrt{5})$ and conjugate to each other. If $x \in 5A$, then $x^{-1} \in 5A$, but $x^2, x^3 \in 5B$; these conjugacy classes behave like the ones of the same order in A_5.

Proposition 7.4.6 (cf. [Ho]) *The triple $(2A, 5A, 5B)$ is rigid but not strictly rigid.*

The relevant part of the character table of $G = J_1$ is (cf. [ATLAS], p.36):

	175560	120	30	30	orders of ←centralizers
	$1A$	$2A$	$5A$	$5B$	← classes
χ_1	1	1	1	1	
χ_4	76	4	1	1	
χ_5	76	-4	1	1	
χ_6	77	5	2	2	$z = (1 + \sqrt{5})/2$
χ_7	77	-3	$-z'$	$-z$	$z' = (1 - \sqrt{5})/2$
χ_8	77	-3	$-z$	$-z'$	
χ_{12}	133	5	-2	-2	
χ_{13}	133	-3	z'	z	
χ_{14}	133	-3	z	z'	
χ_{15}	209	1	-1	-1	

Using formula (7.3) of §7.3, one obtains:

$$
\begin{aligned}
|\bar\Sigma| &= \tfrac{175560^2}{120\cdot30^2}(1+\tfrac{4}{76}-\tfrac{4}{76}+\tfrac{20}{77}+\tfrac{3}{77}+\tfrac{3}{77}+\tfrac{20}{133}+\tfrac{3}{133}+\tfrac{3}{133}+\tfrac{1}{209})\\
&= 438900 = \tfrac{5}{2}|G|.
\end{aligned}
$$

Hence $(2A, 5A, 5B)$ is not strictly rigid. One can check that the triples in $\bar\Sigma - \Sigma$ generate subroups of J_1 isomorphic to A_5. It is known that J_1 contains 2 conjugacy classes of such subgroups:

1. There are $|J_1|/(2|A_5|)$ conjugate subgroups isomorphic to A_5 which are contained in the centralizer of an involution in J_1. (Indeed, J_1 was first defined abstractly by Janko as a simple group having the property that it contains an involution whose centralizer is isomorphic to $\{\pm1\} \times A_5$.)

2. There is a conjugacy class of A_5-subgroups which are self-normalizing: there are $|J_1|/|A_5|$ such subgroups.

In all, one has $\tfrac{3}{2}|J_1|/|A_5|$ subgroups of J_1 which are isomorphic to A_5. Since the conjugacy classes $(2A, 5A, 5B)$ are rigid in A_5, each subgroup gives $|A_5|$ solutions in $\bar\Sigma$. This shows that

$$
|\bar\Sigma - \Sigma| = \frac{3}{2}|J_1|,
$$

and hence $|\Sigma| = |J_1|$, i.e., $(2A, 5A, 5B)$ is rigid.

7.4.6   The Hall-Janko group J_2

This sporadic simple group has order $604800 = 2^7 3^3 5^2 7$. It has a rational class $7A$ of order 7, and two conjugate classes $5A$ and $5B$ of order 5, rational over $\mathbf{Q}(\sqrt{5})$, see e.g. [ATLAS], pp.42-43.

Proposition 7.4.7 (cf. [Ho]) *The triple $(5A, 5B, 7A)$ is strictly rigid.*

No proper subgroup of J_2 has order divisible by 35, hence $|\bar\Sigma| = |\Sigma|$. On the other hand, formula (7.3) of §7.3 and the character table of J_2 ([ATLAS], *loc. cit.*) give:

$$
|\bar\Sigma| = \frac{|J_2|^2}{300^2 \cdot 7}(1 + \frac{16}{36} - \frac{25}{90} - \frac{25}{160} + \frac{9}{288}) = |J_2|.
$$

Hence the result.

7.4.7 The Fischer-Griess Monster M

The Fischer-Griess group M, known as the "Monster", is the largest of the sporadic simple groups. Its order is

$$2^{46} \cdot 3^{20} \cdot 5^9 \cdot 7^6 \cdot 11^2 \cdot 13^3 \cdot 17 \cdot 19 \cdot 23 \cdot 29 \cdot 31 \cdot 41 \cdot 47 \cdot 59 \cdot 71,$$

yet it has only 194 conjugacy classes. Its character table is therefore of manageable size (in fact, it was computed before M had been shown to exist). The group M contains rational conjugacy classes $2A$, $3B$, and $29A$ of exponent 2, 3, and 29 (ATLAS notation).

Proposition 7.4.8 (Thompson, cf. [Hunt], [Ma3], [Th2])
The triple $(2A, 3B, 29A)$ is strictly rigid.

It can be verified by computer that $|\bar{\Sigma}| = |M|$. To prove rigidity, one must show that $\Sigma = \bar{\Sigma}$, i.e., no $(x, y, z) \in \bar{\Sigma}$ generate a proper subgroup of M. Unfortunately, the maximal subgroups of M are not completely known at present. Hence, one must take the following indirect approach: suppose there is a proper subgroup G in M which is generated by $(x, y, z) \in \bar{\Sigma}$. Let S be a simple quotient of G. Clearly the elements x, y, z have non-trivial image in S, and hence the order of S is divisible by $2 \cdot 3 \cdot 29$. Hence, it suffices to check that there are no simple groups S with $2 \cdot 3 \cdot 29$ dividing $|S|$ and $|S|$ dividing $|M|$, such that S is generated by elements x, y, z coming from the conjugacy classes $2A$, $3B$, and $29A$ in the Monster. This is done by checking that no group in the list of finite simple groups satisfies these properties. One is thus forced to invoke the classification theorem for the finite simple groups to prove rigidity in this case.

[Although the proof of the classification theorem had been announced, described and advertised since 1980, it was not complete at the time the present course was given (1988): the part on "quasi-thin" groups was lacking. This gap was only filled in 2004 by M. Aschbacher and S. D. Smith [AS].]

Chapter 8

Construction of Galois extensions of Q(T) by the rigidity method

8.1 The main theorem

Let K be a field of characteristic zero, let $P_1, \ldots, P_k$ be distinct K-rational points of $\mathbf{P}_1$, and let $C_1, \ldots, C_k$ be a family of conjugacy classes of a finite group G with trivial center. The following result is due to Belyi, Fried, Matzat, Shih, and Thompson (see [Ma23], [MM] and [Se8] for references).

Theorem 8.1.1 *Assume that the family* $(C_1, \ldots, C_k)$ *is rigid and that each* C_i *is rational. Then there is a regular G-covering $C \longrightarrow \mathbf{P}_1$ defined over K which is unramified outside $\{P_1, \ldots, P_k\}$ and such that the inertia group over each P_i is generated by an element of C_i. Furthermore, such a covering is unique, up to a unique G-isomorphism.*

By taking $K = \mathbf{Q}$, one has:

Corollary 8.1.2 *G has property* Gal_T *(and hence is a Galois group over* $\mathbf{Q}$*).*

Proof of th. 8.1.1
Let L be the maximal extension of $\bar{K}(T)$ unramified outside $\{P_1, \ldots, P_k\}$, and let π denote the Galois group of L over $\bar{K}(T)$. This is the algebraic fundamental group of $\mathbf{P}_1 - \{P_1, \ldots, P_k\}$ over $\bar{K}$. (It is also called the geometric fundamental group because the ground field is algebraically closed.)

Since L is a Galois extension of $K(T)$, one has an exact sequence:

$$1 \longrightarrow \pi \longrightarrow \pi_K \longrightarrow \Gamma \longrightarrow 1, \tag{8.1}$$

where π_K is the Galois group of L over $K(T)$, and $\Gamma = \mathrm{Gal}(\bar{K}/K)$.

Let I_i be the inertia group of π at P_i. As a profinite group, π has a presentation given by k generators $x_1, \ldots, x_k$ and a single relation $x_1 \cdots x_k = 1$, cf. th. 6.3.1. More precisely, choose a coherent system $\{z_\alpha\}$ of roots of unity in $\bar{\mathbf{Q}}$. This choice determines an element x_i in each I_i up to conjugacy in π. One can then choose the x_i so that they satisfy the relation $x_1 \cdots x_k = 1$ (cf. §6.3).

The set $\mathrm{Hom}(\pi, G)$ of continuous homomorphisms $\pi \longrightarrow G$ is equipped with natural G- and π_K-actions. The G-action is defined (on the left) by

$$(g * f)(x) = g f(x) g^{-1}, \qquad g \in G, \quad f \in \mathrm{Hom}(\pi, G),$$

and the π_K-action is defined (on the right) by

$$(f * \alpha)(x) = f(\alpha x \alpha^{-1}), \qquad \alpha \in \pi_K, \quad f \in \mathrm{Hom}(\pi, G).$$

The two actions commute, i.e.,

$$(g * f) * \alpha = g * (f * \alpha).$$

Consider the set $H \subset \mathrm{Hom}(\pi, G)$ defined by:

$$H = \{\phi | \phi \text{ is surjective and } \phi(x_i) \in C_i \text{ for all } i\}.$$

This set is stable under both the G and π_K-actions:

- The action of G on itself by inner automorphisms stabilizes the C_i, and hence G preserves H, which is isomorphic to $\Sigma(C_1, \ldots, C_k)$ as a G-set (cf. §7.3). By the rigidity assumption, G acts freely and transitively on H.

- Conjugation by an element $\sigma \in \pi_K$ sends an inertia group I_i at P_i to an inertia group at P_i^σ. Since the P_i are K-rational, π_K permutes the inertia groups above P_i. Hence, σ sends each of the $x_i \in I_i$ to a conjugate of x_i^α, for some $\alpha \in \hat{\mathbf{Z}}$ (namely, $\alpha = \chi(\sigma)$, where χ is the cyclotomic character). By the rationality of the C_i, it follows that $f * \sigma$ maps each x_i to an element of C_i, and hence $f * \sigma \in H$.

Any $\phi \in H$ defines a Galois extension E of $\bar{K}(T)$ with Galois group G. To descend from $\bar{K}(T)$ to $K(T)$, it suffices to show that ϕ can be extended to π_K. This is an immediate consequence of the following:

Lemma 8.1.3 *Let* $1 \longrightarrow A \longrightarrow B \longrightarrow C \longrightarrow 1$ *be an exact sequence of groups, let G be a finite group, and let G and B act on $\mathrm{Hom}(A, G)$ as above (i.e., $(g * f)(x) = gf(x)g^{-1}$ if $g \in G$, $x \in A$, and $(f * b)(x) = f(bxb^{-1})$ if $b \in B$, $x \in A$). If H is a non-empty subset of $\mathrm{Hom}(A, G)$ on which G acts freely and transitively, then the following are equivalent:*

1. *Any $\phi \in H$ extends uniquely to a homomorphism $B \longrightarrow G$.*
2. *H is stable under the action of B.*

Proof: Suppose 1 holds, i.e., any $\phi \in H$ extends uniquely to a homomorphism $\psi : B \longrightarrow G$. If $b \in B$, then $(\phi * b)(x) = \phi(bxb^{-1}) = \psi(bxb^{-1}) = (\psi(b) * \phi)(x)$. Hence $\phi * b \in H$, since by hypothesis H is preserved by the action of G. Conversely, if property 2 is satisfied, then given $\phi : A \longrightarrow G$, one may define $\psi : B \longrightarrow G$ by:

$$\phi * b = \psi(b) * \phi.$$

Such a ψ exists (G acts transitively on H) and is unique (G acts freely on H). One verifies that ψ defines a homomorphism $B \longrightarrow G$ which extends ϕ; this follows from the compatibility of the G- and B-actions.

This completes the proof of the lemma, and hence of th. 8.1.1.

Alternate method of proof for th. 8.1.1:
(a) Prove that the required G-covering $C \longrightarrow \mathbf{P}_1$ exists over $\bar{K}$ and is unique up to a unique isomorphism.
(b) Use Weil's descent criterion ([Se3], chap. V, no. 20) to prove that C, together with the action of G, can be defined over K. (General principle: every "problem" over K which has a unique solution - up to a unique isomorphism - over $\bar{K}$ has a solution over K.)

Remark: When $k = 3$, one can suppose without loss of generality that $(P_1, P_2, P_3) = (0, 1, \infty)$. In this way, a rigid triple (C_1, C_2, C_3) of rational conjugacy classes of G determines a canonical extension of $K(T)$. Several natural questions arise in this context:

1. Can one describe what happens when K is a local field? We will do this (in a special case) for $K = \mathbf{R}$ in §8.4.

2. Can one describe the decomposition group above P_i? For example, if G is the Monster, and $(C_1, C_2, C_3) = (2A, 3B, 29A)$ as in §7.4.7, then the decomposition group D_1 above P_1 must be contained in the normalizer of an element of the class $2A$. This normalizer is $2 \times B$, where B is the "Baby Monster" sporadic group. Aside from this, nothing seems to be known about D_1.

Exercise: Show that the alternating group A_n ($n = 4, 5, 6, 7, 8$) does not contain any rigid family $(C_1, \ldots, C_k)$ of rational conjugacy classes (use remark 1 of §7.3).

8.2   Two variants

8.2.1   First variant

Th. 8.1.1 can be generalized to the case where the classes are only K-rational. More precisely, let us fix a choice of primitive Nth roots of unity over K, i.e., an orbit under $\mathrm{Gal}(\bar{K}/K)$ of primitive Nth roots of unity. This amounts to choosing a K-irreducible factor of the Nth cyclotomic polynomial ϕ_N. (For example, if $K = \mathbf{Q}(\sqrt{5})$, $N = 5$, the cyclotomic polynomial ϕ_5 factors as

$$\phi_5(X) = X^4 + X^3 + X^2 + X + 1 = (X^2 + \frac{1 + \sqrt{5}}{2}X + 1)(X^2 + \frac{1 - \sqrt{5}}{2}X + 1).$$

A choice of 5th roots of unity gives a square root of 5 in $\mathbf{Q}(\sqrt{5})$.) Such a choice determines a generator x_i for each inertia group I_i at P_i (which is well-defined up to conjugation in π_K). After such a choice has been made, one has:

Theorem 8.2.1 *If $C_1, \ldots, C_k$ is a rigid family of K-rational classes of G, and P_1, ..., P_k are K-rational points of $\mathbf{P}_1$, then there is a regular G-covering $C \longrightarrow \mathbf{P}_1$ defined over K which is unramified outside $\{P_1, \ldots, P_k\}$ and such that the x_i-generator of the inertia group above P_i belongs to the class C_i. This covering is uniquely defined up to a unique G-isomorphism.*

The proof is essentially the same as that of th. 8.1.1.

Since conjugacy classes are always rational over the maximal cyclotomic extension $\mathbf{Q}^{\mathrm{cycl}}$ of $\mathbf{Q}$, one only needs the rigidity condition to ensure that a group can be realized as a Galois group over $\mathbf{Q}^{\mathrm{cycl}}(T)$. This property is known for:

- Most of the classical Chevalley groups over finite fields (Belyi [Be2]);

- All the sporadic groups;

- Most of the exceptional groups G_2, F_4, E_6, E_7 and E_8 (and also the twisted forms $^2G_2$, $^3D_4$, and $^2E_6$) over finite fields (Malle [Ml1]).

8.2.2   Second variant

The assumption that the conjugacy classes are rational is often too restrictive for applications. The following variant of th. 8.1.1 is useful in practice:

Theorem 8.2.2 *Let (C_1, C_2, C_3) be a rigid triple of conjugacy classes of G, with C_1 rational and C_2 and C_3 conjugate to each other over a quadratic field $\mathbf{Q}(\sqrt{D})$. Let $P_1 \in \mathbf{P}_1(\mathbf{Q})$, $P_2, P_3 \in \mathbf{P}_1(\mathbf{Q}(\sqrt{D}))$, with P_2 and P_3 conjugate to each other. Then there is a regular G-extension of $\mathbf{Q}(T)$ which is ramified only at P_1, P_2 and P_3, and such that the canonical generator of the inertia group at P_i (which is well-defined after a choice of roots of unity over $\mathbf{Q}(\sqrt{D})$) is in C_i.*

Corollary 8.2.3 *The group G has property Gal_T.*

The proof of th. 8.2.2 is similar to that of th. 8.1.1. The set $H \subset \mathrm{Hom}(\pi, G)$ is defined in the same way. The key point is to prove that H is still preserved under the action of $\pi_{\mathbf{Q}}$.

- If $\sigma \in \pi_{\mathbf{Q}}$ is trivial on $\mathbf{Q}(\sqrt{D})$, then σ fixes P_1, P_2 and P_3, and also fixes the choice of roots of unity over $\mathbf{Q}(\sqrt{D})$. Hence σ preserves H, as before.

- If σ is not trivial on $\mathbf{Q}(\sqrt{D})$, then σ interchanges P_2 and P_3, and hence I_2 and I_3. But σ also changes the choice of roots of unity, and these effects compensate each other.

Remarks:
1. The assumptions on the number of classes and the field of rationality are only put to simplify the proof, and because this is the principal case which occurs in practice. In fact, the same conclusion holds in greater generality, e.g., if $\{C_1, \ldots, C_k\}$ is stable under the action of $\mathrm{Gal}(\bar{\mathbf{Q}}/\mathbf{Q})$, and the map $\{P_1, \ldots, P_k\} \longrightarrow \{C_1, \ldots, C_k\}$ defined by $P_i \mapsto C_i$ is an anti-isomorphism of $\mathrm{Gal}(\bar{\mathbf{Q}}/\mathbf{Q})$-sets.
2. For other variants of th. 8.1.1 using the braid group, see [Fr1] and [MM].

8.3 Examples

Here also, we only give a few examples. For more, see [MM].

8.3.1 The symmetric group S_n

Recall from §7.4.1 that the symmetric group S_n has a rigid triple of conjugacy classes $(nA, 2A, C^{(k)})$, when $(k, n) = 1$. The covering $\mathbf{P}_1 \longrightarrow \mathbf{P}_1$ given by $X \mapsto X^k (X-1)^{n-k}$ has ramification of this type, namely:

$$
\begin{cases}
t = \infty & nA \\
t = 0 & C^{(k)} \\
t = k^k (k-n)^{n-k} n^{-n} & 2A
\end{cases}
$$

Hence, by th. 8.1.1, the polynomial

$$X^k(X-1)^{n-k} - T = 0$$

has Galois group S_n over $\mathbf{Q}(T)$ when $(k,n) = 1$. Note that when $(k,n) = l \neq 1$, the splitting field of the equation

$$X^k(X-1)^{n-k} - T = 0$$

contains $\mathbf{Q}(T^{1/l}, \mu_l)$. Its Galois group is strictly smaller than S_n (and the extension is not regular when $l > 2$).

8.3.2   The alternating group A_5

Recall from §7.4.2 that the triple $(2A, 3A, 5A)$ of conjugacy classes in A_5 is rigid. The conjugacy class $5A$ is rational over $\mathbf{Q}(\sqrt{5})$ (but not over $\mathbf{Q}$). By th. 8.2.1, there is a regular extension of $\mathbf{Q}(\sqrt{5})(T)$ with Galois group A_5, and ramification of type $(2A, 3A, 5A)$. The corresponding curve C has genus 0 (but is not isomorphic to $\mathbf{P}_1$ over $\mathbf{Q}(\sqrt{5})$, cf. [Se5]).

   The action of A_5 on C can be realized geometrically as follows (loc. cit.). Consider the variety in $\mathbf{P}_4$ defined by the equations:

$$\begin{cases} X_1 + \cdots + X_5 = 0 \\ X_1^2 + \cdots + X_5^2 = 0. \end{cases}$$

Since the first equation is linear, this variety can be viewed as a quadric hypersurface in $\mathbf{P}_3$. The variety of lines on this quadric is a curve over $\mathbf{Q}$ which becomes isomorphic over $\mathbf{Q}(\sqrt{5})$ to the disjoint union of two curves of genus 0 which are conjugate over $\mathbf{Q}(\sqrt{5})$. The obvious action of S_5 on V (permuting coordinates) induces an action of S_5 on this curve. The extension of $\mathbf{Q}(T)$ corresponding to this curve is a non-regular extension with Galois group S_5, which contains $\mathbf{Q}(\sqrt{5})$; it can also be viewed as a regular A_5-extension of $\mathbf{Q}(\sqrt{5})(T)$.

   An A_5-covering of $\mathbf{P}_1$ with ramification $(2A, 5A, 5B)$ can be realized by taking an S_5-covering with ramification of type $(2, 4, 5)$ (e.g., the covering $X \mapsto X^5 - X^4$) and using the double group trick (cf. §4.5). This defines a regular A_5-covering of $\mathbf{P}_1$ over $\mathbf{Q}$, with two ramification points conjugate over $\mathbf{Q}(\sqrt{5})$; this is the situation of th. 8.2.2. This covering can be shown to be isomorphic to the Bring curve, defined in $\mathbf{P}_4$ by the homogeneous equations

$$\begin{cases} X_1 + \cdots + X_5 = 0, \\ X_1^2 + \cdots + X_5^2 = 0, \\ X_1^3 + \cdots + X_5^3 = 0, \end{cases}$$

cf. §4.4, exercise.

Exercise: Let C be the Bring curve in $\mathbf{P}_4$ (see above).
a) Let E be the quotient of C by the group of order 2 generated by the transposition (12) in S_5. Show that E is isomorphic to the elliptic curve defined in $\mathbf{P}_2$ by the homogeneous equation:

$$(x^3 + y^3 + z^3) + (x^2y + x^2z + y^2z + y^2x + z^2x + z^2y) + xyz = 0,$$

(put $x = x_3$, $y = x_4 = z = x_5$). Show that this curve is $\mathbf{Q}$-isomorphic to the curve $50E$ of [ANVERS], p.86, with minimal equation

$$Y^2 + XY + Y = X^3 - X - 2$$

and j-invariant $-5^2/2$.
b) Use the action of S_5 on C to show that the Jacobian of C is $\mathbf{Q}$-isogenous to the product of 4 copies of E.
c) Show that the quotient of C by A_4 is $\mathbf{Q}$-isomorphic to the elliptic curve $50H$ of [ANVERS], *loc.cit.*, with j-invariant $2^{-15} \cdot 5 \cdot 211^3$; this curve is 15-isogenous to E.
d) Show that C has good reduction (mod p) for $p \neq 2, 5$. If $N_p(C)$ (resp $N_p(E)$) denotes the number of points on C (resp. E) modulo p, deduce from b) that

$$N_p(C) = 4N_p(E) - 3 - 3p.$$

Use [ANVERS], p.117, to construct a table giving $N_p(C)$ for $p < 100$:

p	7	11	13	...	83	89	97
$N_p(C)$	0	24	30	...	120	30	90

Check these values by determining the polynomials $X^5 + aX + b$ over $\mathbf{F}_p$, with $(a,b) \neq (0,0)$, which have all their roots rational over $\mathbf{F}_p$. For instance, if $p = 83$, there is only one such polynomial (up to replacing a by at^4 and b by bt^5, with $t \in \mathbf{F}_p^*$), namely:

$$X^5 + 11X + 11 \equiv (X + 33)(X + 13)(X - 4)(X - 20)(X - 22).$$

This fits with $N_p(C) = 120$.
e) Show that C has semi-stable reduction at 2, the reduced curve being isomorphic (over $\mathbf{F}_4$) with the union of two copies of $\mathbf{P}_1$ intersecting each other at the five points of $\mathbf{P}_1(\mathbf{F}_4)$; describe the action of S_5 on this curve, using the fact that A_5 is isomorphic to $\mathbf{SL}_2(\mathbf{F}_4)$.

8.3.3   The group $\mathbf{PSL}_2(\mathbf{F_p})$

Th. 8.2.2 applied to the rigid triples of conjugacy classes $(2A, pA, pB)$ (when $\left(\frac{2}{p}\right) = -1$) and $(3A, pA, pB)$ (when $\left(\frac{3}{p}\right) = -1$) shows that there are regular $\mathbf{PSL}_2(\mathbf{F}_p)$-extensions of $\mathbf{Q}(T)$ with ramification of this shape. These correspond to the Shih coverings with $N = 2$ and $N = 3$ (cf. §5.1).

The rigidity method does not predict the existence of the Shih coverings related to $N = 7$, which are ramified at four points.

A covering of $\mathbf{P}_1$ having ramification type $(2A, 3A, pA)$ is given by the covering of modular curves defined over $\mathbf{Q}(\sqrt{p^*})$, $X(p) \longrightarrow X(1)$. This covering has Galois group $\mathbf{PSL}_2(\mathbf{F}_p)$ over $\mathbf{Q}(\sqrt{p^*})$; the rigidity property shows that it is the only $\mathbf{PSL}_2(\mathbf{F}_p)$-covering with this ramification type. This was first pointed out by Hecke [He].

In particular, the rigid triple $(2A, 3A, 7A)$ in $\mathbf{PSL}_2(\mathbf{F}_7)$ gives rise to the Klein covering of $\mathbf{P}_1$ defined over $\mathbf{Q}(\sqrt{-7})$ and having Galois group $\mathbf{PSL}_2(\mathbf{F}_7)$. Its function field E is a Galois extension of $\mathbf{Q}(T)$ with Galois group $\mathbf{PGL}_2(\mathbf{F}_7)$, which is not regular.

8.3.4   The Gal_T property for the smallest simple groups

The following table lists the smallest ten non-abelian simple groups. All of them, except the last one $\mathbf{SL}_2(\mathbf{F}_{16})$, are known to have the Gal_T property. The last column of the table indicates why.

group	order	Gal_T property
$A_5 = \mathbf{SL}_2(\mathbf{F}_4)$	$60 = 2^2 \cdot 3 \cdot 5$	Hilbert (§§4.5, 8.3.2)
$\quad = \mathbf{PSL}_2(\mathbf{F}_5)$		Shih with $N = 2, 3$
		(§§5.1, 8.3.3);
$\mathbf{SL}_3(\mathbf{F}_2)$	$168 = 2^3 \cdot 3 \cdot 7$	Shih with $N = 3$
$\quad = \mathbf{PSL}_2(\mathbf{F}_7)$		(§§5.1, 8.3.3);
$A_6 = \mathbf{PSL}_2(\mathbf{F}_9)$	$360 = 2^3 \cdot 3^2 \cdot 5$	Hilbert (§4.5);
$\mathbf{SL}_2(\mathbf{F}_8)$	$504 = 2^3 \cdot 3^2 \cdot 7$	Th. 8.2.2 and §7.4.4;
$\mathbf{PSL}_2(\mathbf{F}_{11})$	$660 = 2^2 \cdot 3 \cdot 5 \cdot 11$	Shih with $N = 2$
		(§§5.1, 8.3.3);
$\mathbf{PSL}_2(\mathbf{F}_{13})$	$1092 = 2^2 \cdot 3 \cdot 7 \cdot 13$	Shih with $N = 2$
		(§§5.1, 8.3.3);
$\mathbf{PSL}_2(\mathbf{F}_{17})$	$2448 = 2^4 \cdot 3^2 \cdot 17$	Shih with $N = 3$
		(§5.1, 8.3.3);
A_7	$2520 = 2^3 \cdot 3^2 \cdot 5 \cdot 7$	Hilbert (§4.5);
$\mathbf{PSL}_2(\mathbf{F}_{19})$	$3420 = 2^2 \cdot 3^2 \cdot 5 \cdot 19$	Shih with $N = 2, 3$
		(§§5.1, 8.3.3);
$\mathbf{SL}_2(\mathbf{F}_{16})$	$4080 = 2^4 \cdot 3 \cdot 5 \cdot 17$	?

In addition, it has been shown that all the sporadic simple groups satisfy the Gal_T property, with the possible exception of the Mathieu group M_{23}, cf. [Hunt], [Ma3], [MM], [Pa].

8.4   Local properties

8.4.1   Preliminaries

Let $\pi : C \longrightarrow \mathbf{P}_1$ be a G-covering defined over a field K which is complete with respect to a real valuation. Let x be a point of $\mathbf{P}_1(K) - S$, where S is the ramification locus of the covering. The fiber Λ_x at x is an étale K-algebra with action of G; its structure is defined by a continuous homomorphism

$$\phi_x : \mathrm{Gal}(\bar{K}/K) \longrightarrow G,$$

which is well-defined up to inner conjugation in G. Let us denote by H the quotient of the set of such homomorphisms by the action of G by conjugation; we endow H with the discrete topology.

Proposition 8.4.1 *The map from $\mathbf{P}_1(K)$ to H defined by $x \mapsto \phi_x$ is continuous with respect to the topology of $\mathbf{P}_1(K)$ induced by the valuation on K. (In other words, for every $x \in \mathbf{P}_1(K)$, there is a neigbourhood U of x such that the Galois algebras Λ_x and Λ_y are isomorphic, for all $y \in U$.)*

This can be deduced (with some care) from Krasner's lemma: if

$$P(X) = \prod(X - \alpha_i) \text{ and } Q(X) = \prod(X - \alpha_i)$$

are irreducible polynomials over K, with α_i and β_i sufficiently close for each i, then $K(\alpha_1) = K(\beta_1)$. See [Sa1].

Another way to prove prop. 8.4.1 is as follows: the map $C \longrightarrow \mathbf{P}_1$ gives an extension $\mathcal{O}_x \subset \tilde{\mathcal{O}}_x$, where $\mathcal{O}_x$ is the local ring at x, and $\tilde{\mathcal{O}}_x$ is the semi-local ring above x. Let $\mathcal{M}_x$ be the maximal ideal of $\mathcal{O}_x$. The quotient $\tilde{\mathcal{O}}_x/\mathcal{M}_x\tilde{\mathcal{O}}_x$ is Λ_x. One has a natural inclusion $\mathcal{O}_x \subset \mathcal{O}_x^{\mathrm{hol}}$, where $\mathcal{O}_x^{\mathrm{hol}}$ denotes the ring of power series with coefficients in K which converge in a neighbourhood of x. Since $\mathcal{O}_x^{\mathrm{hol}}$ is Henselian (cf. [Ra1], p. 79), and the extension is étale, one has:

$$\tilde{\mathcal{O}}_x^{\mathrm{hol}} \simeq \Lambda_x \otimes_K \mathcal{O}_x^{\mathrm{hol}}.$$

This means that, for its analytic structure, the covering $C \longrightarrow \mathbf{P}_1$ is locally a product. The proposition follows.

By the discussion above, we may write $\mathbf{P}_1(K) - S$ as a finite disjoint union of open sets:

$$\mathbf{P}_1(K) - S = \bigcup_{\phi \in H} U_\phi,$$

where $U_\phi = \{t \in \mathbf{P}_1(K) - S | \phi_t = \phi\}$. For example, if $K = \mathbf{Q}_p$, $p \neq 2$, the universal $\mathbf{Z}/2\mathbf{Z}$-covering defined by $x \mapsto x^2$ gives a decomposition of

$\mathbf{P}_1(\mathbf{Q}_p) - \{0, \infty\}$ into four open and closed pieces, corresponding to the four non-isomorphic Galois algebras of rank 2 over $\mathbf{Q}_p$.

Remark: A propos of the U_ϕ, let us mention an unpublished result of Raynaud: if the order of G is prime to the residue characteristic, every unramified G-covering of a rigid polydisk becomes trivial after a finite extension of the local field K.

8.4.2   A problem on good reduction

Assume that $\pi : C \longrightarrow \mathbf{P}_1$ is a G-covering obtained by the rigidity method (th. 8.1.1) from rational points $P_1, \ldots, P_k$ and conjugacy class $C_1, \ldots, C_k$.

Let p be a prime number. Assume the reductions of the P_i are distinct in $\mathbf{P}_1(\mathbf{F}_p)$. The following problem was raised in the original Harvard lectures:

Problem: *Assume that p does not divide the orders of the elements of C_i for $i = 1, \ldots, k$. Is it true that the curve C has good reduction at p?*

This is not always true: B. Matzat has pointed out a counterexample with $G = \mathrm{Aut}(S_6), k = p = 3$, cf. J. Crelle 349 (1984), p. 215. But it is true when p does not divide $|G|$, cf. [Bc].

Exercise: Check that the above problem has a positive answer for the examples given in 8.3.1, 8.3.2, 8.3.3.

8.4.3   The real case

We now restrict our attention to the case when $\pi : Y \longrightarrow \mathbf{P}_1$ comes from a rigid triple of conjugacy classes (C_1, C_2, C_3), and where $K = \mathbf{R}$. By prop. 8.4.1, to each connected component of $\mathbf{P}_1(\mathbf{R}) - S$ there is attached a unique conjugacy class of involutions in G, corresponding to complex conjugation. There are two cases, depending on the number of such connected components:

Case 1: We suppose that the three conjugacy classes are rational over $\mathbf{R}$. The corresponding covering is ramified at exactly three real points P_1, P_2, P_3, which divide the circle $\mathbf{P}_1(\mathbf{R})$ into three connected components.

Let us choose $x_i \in C_i$, with $x_1 x_2 x_3 = 1$ and $G = \langle x_1, x_2, x_3 \rangle$. We assume $G \neq \{1\}$, hence $x_i \neq 1$ for $i = 1, 2, 3$. Since the C_i are rational over $\mathbf{R}$, we have

$$x_1^{-1} \in C_1, \qquad x_2^{-1} \in C_2.$$

Put $x_3' = x_2 x_1 = x_2 x_3 x_2^{-1}$. Then $(x_1^{-1}, x_2^{-1}, x_3')$ is in Σ. By rigidity, there exists a unique $s_3 \in G$ such that:

$$\begin{cases} s_3 x_1 s_3^{-1} &= x_1^{-1} \\ s_3 x_2 s_3^{-1} &= x_2^{-1} \end{cases}$$

The element s_3 thus defined is such that $s_3^2 = 1$.

Theorem 8.4.2 *The complex conjugation attached to a point x in the connected component (P_1, P_2) of $\mathbf{P}_1(\mathbf{R})$ between P_1 and P_2 is in the conjugacy class of s_3.*

Remark: An analogous statement holds for the involutions s_1 and s_2, which are defined in the same way as s_3, and correspond to the complex conjugations attached to the components (P_2, P_3) and (P_1, P_3) respectively. This gives the relations:

$$\begin{cases} s_1 &= s_3 x_2, \\ s_2 &= s_1 x_3, \\ s_3 &= s_2 x_1. \end{cases}$$

Hence, if s_1 and s_3 are non-trivial, the group generated by s_1, s_3 and x_2 is a dihedral group of order $2n$, where n is the order of x_2. There is a single conjugacy class of involutions in a dihedral group of order $2n$ with n odd. Hence, one has:

Corollary 8.4.3 *If the order of x_2 is odd, then s_1 and s_3 are conjugate in G.*

Corollary 8.4.4 *If two of the three classes C_1, C_2 and C_3 have odd exponent, then the s_i's are conjugate.*

Example: The rigid triple of conjugacy classes $(2A, 3B, 29A)$ in the Monster M satisfies the hypotheses of cor. 8.4.4. Hence, complex conjugation corresponds to a unique class of involutions in M (the class $2B$, in ATLAS notation).

Remark: The case where $s_3 = 1$ occurs only in the case $G = D_n$ and $(C_1, C_2, C_3) = (2, 2, n)$, n odd. For, if $s_3 = 1$, then x_1, x_2 are of order 2, equal to s_2 and s_1 respectively. The group generated by x_1, x_2, x_3 is thus a dihedral group of order $2n$, where n is odd (since the group has no center). The only dihedral group of this type satisfying the rationality condition over $\mathbf{Q}$ is $G = S_3$. This means that the rigidity method of th. 8.1.1, when applied with three classes to a group $G \neq S_3$, *never gives totally real extensions* of $\mathbf{Q}$.

Case 2: C_1 is rational over $\mathbf{R}$, and C_2 and C_3 are complex conjugate to each other.

Then, $\mathbf{P}_1(\mathbf{R}) - \{P_1, P_2, P_3\}$ is connected, since P_2 and P_3 do not lie on $\mathbf{P}_1(\mathbf{R})$. Hence there is a single conjugacy class of involutions in G corresponding to complex conjugation.

Let $(x_1, x_2, x_3) \in \Sigma(C_1, C_2, C_3)$. By rigidity, there is a unique involution s in G such that:

$$sx_1s^{-1} = x_2^{-1}, \qquad sx_2s^{-1} = x_1^{-1}, \qquad sx_3s^{-1} = x_3^{-1}.$$

Theorem 8.4.5 *Complex conjugation belongs to the conjugacy class of s.*

Observe that in this case $s \neq 1$. (Otherwise G would be cyclic.)

Proof of th. 8.4.2 and th. 8.4.5 Choose a base point x on $\mathbf{P}_1(\mathbf{R})$, (lying on the connected component (P_1, P_2) in case 1) and let π denote as before the geometric fundamental group of $X = \mathbf{P}_1 - \{P_1, P_2, P_3\}$, with base point x. Let $\alpha_1, \alpha_2, \alpha_3$ denote the generators of π corresponding to paths around P_1, P_2, and P_3 respectively. The complex conjugation σ_X on X acts as a symmetry around the equator on the Riemann sphere. Hence σ_X acts on the generators $\alpha_1, \alpha_2, \alpha_3$ of π by:
Case 1: $\alpha_1 \mapsto \alpha_1^{-1}$, $\quad \alpha_2 \mapsto \alpha_2^{-1}$
Case 2: $\alpha_2 \mapsto \alpha_3^{-1}$, $\quad \alpha_3 \mapsto \alpha_2^{-1}$

Let σ_Y denote the complex conjugation acting on Y. Since the G-covering $\pi : Y \longrightarrow X$ is defined over $\mathbf{R}$, the following diagram commutes:

$$
\begin{array}{ccc}
Y & \xrightarrow{\sigma_Y} & Y \\
\pi \downarrow & & \downarrow \pi \\
X & \xrightarrow{\sigma_X} & X
\end{array}
$$

Since $\sigma_X x = x$, the fiber Y_x of Y over x is preserved by σ_Y. The action of σ_Y on Y_x commutes with the action of G.

The group G acts freely and transitively on Y_x, so a choice of $y \in Y_x$ determines a surjective map $\phi_y : \pi_1(X; x) \longrightarrow G$. Because the action of σ commutes with that of G, the following diagram is commutative:

$$
\begin{array}{ccc}
\pi_1(X; x) & \xrightarrow{\phi_y} & G \\
\sigma_X \downarrow & & \| \\
\pi_1(X; x) & \xrightarrow{\phi_{\sigma y}} & G
\end{array}
$$

On the other hand, we also have:

$$\phi_{\sigma y} = \sigma_Y \phi_y \sigma_Y^{-1}.$$

Hence the diagram below commutes:

$$
\begin{array}{ccc}
\pi_1(X;x) & \xrightarrow{\phi_y} & G \\
\sigma_X \downarrow & & \downarrow \mathrm{Inn}(\sigma_{Y,t}) \\
\pi_1(X;x) & \xrightarrow{\phi_{\sigma y}} & G.
\end{array}
$$

This proves the theorem.

Exercise: Let G be a finite group, generated by elements $g_1, \ldots, g_k$. Choose points $z_1, \ldots, z_k$ with (say) $\mathrm{Re}(z_i) = 0$ for $i = 1, \ldots k$, and

$$\mathrm{Im}(z_1) > \mathrm{Im}(z_2) > \ldots > \mathrm{Im}(z_k) > 0.$$

Let $z_1', \ldots, z_k'$ be the complex conjugates of $z_1, \ldots, z_k$ and choose a real base point $x < 0$; let π denote the fundamental group $\pi_1(\mathbf{P}_1(\mathbf{C}) - S; x)$, where $S = \{z_1, \ldots, z_k, z_1', \ldots, z_k'\}$. It is generated by elements $x_1, \ldots, x_k, x_1', \ldots, x_k'$, where x_i (resp. x_i') denotes the homotopy class of paths going in a straight line from x to z_i (resp. z_i') and going around this point in the positive direction. These generators, together with the relation $x_1 \cdots x_k x_k' \cdots x_1' = 1$, give a presentation for π. Let ϕ be the homomorphism $\pi \longrightarrow G$ sending x_i to g_i and x_i' to g_i^{-1}. Show that ϕ defines a regular G-extension of $\mathbf{R}(T)$. (Hence every finite group is the Galois group of a regular extension of $\mathbf{R}(T)$. This result can also be deduced from [KN]. The analogous statement for $\mathbf{Q}_p(T)$ is also true; see next section.)

8.4.4 The p-adic case: a theorem of Harbater

Let p be a prime number.

Theorem 8.4.6 ([Harb]) *Every finite group is the Galois group of a regular extension of* $\mathbf{Q}_p(T)$.

The proof shows more, namely that for every finite group G, there exists an absolutely irreducible G-covering $X \longrightarrow \mathbf{P}_1$ over $\mathbf{Q}_p$ having a "base point", i.e., an unramified point $P \in \mathbf{P}_1(\mathbf{Q}_p)$ which is the image of a $\mathbf{Q}_p$-point of X. Call R_p this property of G. The theorem clearly follows from the following two assertions:

(i) Every cyclic group has property R_p.

(ii) If G is generated by two subgroups G_1 and G_2 having property R_p, then G has property R_p.

Assertion (i) is easy (and true over $\mathbf{Q}$, as the construction of §4.2 shows). Assertion (ii) is proved by a gluing process which uses rigid analytic geometry. Namely, let $X_i \longrightarrow \mathbf{P}_1$ ($i = 1, 2$) be a G_i-covering as above with a base point P_i. By removing a small neighborhood of P_i one gets a rigid analytic G_i-covering $V_i \longrightarrow U_i$ where U_i is a p-adic disk; this covering has the further property that it is trivial on $U_i - U_i'$ where U_i' is a smaller disk contained in U_i. Let $W_i \longrightarrow U_i$ be the (non-connected) G-covering of U_i defined by induction from G_i to G. (It is a disjoint union of $|G/G_i|$ copies of V_i.) One then embeds U_1 and U_2 as disjoint disks in $\mathbf{P}_1$, and defines a rigid analytic G-covering of $\mathbf{P}_1$ by gluing together W_1 on U_1, W_2 on U_2, and the trivial G-covering on $\mathbf{P}_1 - U_1' - U_2'$. If this is done properly, the resulting G-covering W is absolutely irreducible. By the "GAGA" theorem in the rigid analytic setting (cf. [Ki], [Kö]), this covering is algebraic, and (ii) follows.

Remark: Harbater's original proof uses "formal GAGA" instead of "rigid GAGA"; the idea is the same.

Chapter 9

The form $\text{Tr}(x^2)$ and its applications

9.1 Preliminaries

9.1.1 Galois cohomology (mod 2)

Let K be a field of characteristic $\neq 2$, and let G_K be $\text{Gal}(K_s/K)$, where K_s is a separable closure of K. We will be interested in the Galois cohomology of K modulo 2; for brevity, let us denote the cohomology groups $H^i(G_K, \mathbf{Z}/2\mathbf{Z})$ by $H^i(G_K)$. In the case where $i = 1, 2$, Kummer theory provides the following interpretation of $H^i(G_K)$:

$$H^1(G_K) = K^*/K^{*2},$$

$$H^2(G_K) = \text{Br}_2(K),$$

where $\text{Br}_2(K)$ is the 2-torsion in the Brauer group of K. If $a \in K^*$, we denote by (a) the corresponding element of $H^1(G_K)$. The cup-product

$$(a)(b) \in H^2(G_K)$$

corresponds to the class of the quaternion algebra (a, b) in $\text{Br}_2(K)$ defined by

$$i^2 = a, \quad j^2 = b, \quad ij = -ji.$$

9.1.2   Quadratic forms

Let f be a non-degenerate quadratic form over K of rank $n \geq 1$. By choosing an appropriate basis, we may write

$$f = \sum_{i=1}^{n} a_i X_i^2, \quad \text{with } a_i \in K^*.$$

The element $(1 + (a_1))(1 + (a_2)) \cdots (1 + (a_n))$ in the cohomology ring $H^*(G_K)$ depends only on f. One defines the i-th Stiefel-Whitney class w_i of f by:

$$(1 + (a_1)) \cdots (1 + (a_n)) = 1 + w_1 + w_2 + \cdots + w_n, \quad w_i \in H^i(G_K).$$

In particular, we have:

$$w_1 = \sum (a_i) = (\prod a_i) = (\mathrm{Disc}(f))$$

$$w_2 = \sum_{i<j} (a_i)(a_j).$$

The cohomology class w_2 is known as the Hasse (or Witt) invariant of the quadratic form f. If K is a number field, then f is completely characterized by its rank, signature, and the invariants w_1 and w_2. (The same is true for arbitrary K, when $n \leq 3$.) The following results can be found in [Sch], pp. 211-216.

Theorem 9.1.1 (Springer) *If two quadratic forms over K become equivalent over an odd-degree extension K' of K, then they are already equivalent over K.*

(For a generalization to hermitian forms, see [BL].)

Let us now consider $K(T)$, where T is an indeterminate. If v is a place of $K(T)$ which is trivial on K and $\neq \infty$, there is a unique uniformizing parameter π_v at v which is monic and irreducible in $K[T]$. Let $K(v) = K[T]/(\pi_v)$ denote the residue field at v, and let $a \mapsto \bar{a}$ be the reduction map $K[T] \longrightarrow K(v)$. If $f = \sum_{i=1}^{n} a_i X_i^2$ is a quadratic form over $K(T)$, we may assume (since the a_i can be modified by squares) that $v(a_i) = 0$ or 1. The $K(v)$-quadratic forms:

$$\partial_1(f) = \sum_{v(a_i)=0} \bar{a_i}\, X_i^2, \qquad \partial_2(f) = \sum_{v(a_i)=1} \overline{(a_i/\pi_v)}\, X_i^2$$

are called the first and second *residues* of f. One shows (cf. e.g. [Sch], p. 209) that their images in the *Witt group* $W(K(v))$ do not depend on the chosen representation of f as $\sum a_i X_i^2$. (Recall that the Witt group $W(L)$ of a field L is the Grothendieck group of the set of quadratic forms over L, with the hyperbolic forms identified to 0.)

Theorem 9.1.2 (Milnor) *If a quadratic form f over $K(T)$ has second residue 0 at all places of $K(T)$ except ∞, then f is equivalent to a quadratic form over K.*

(More precisely, one has an exact sequence:

$$0 \longrightarrow W(K) \longrightarrow W(K(T)) \longrightarrow \coprod_{v \neq \infty} W(K(v)) \longrightarrow 0.)$$

A quadratic form over a ring R is said to be *strictly non-degenerate* if its discriminant is invertible in R.

Theorem 9.1.3 (Harder) *A strictly non-degenerate quadratic form over the ring $K[T]$ comes from K.*

This theorem can be formulated more suggestively: a quadratic vector bundle over $\mathbf{A}^1$ is constant. (For a generalization to other types of bundles, see [RR].)

9.1.3 Cohomology of S_n

We need only $H^i(S_n)$ for $i = 1, 2$. These groups are well-known:

$$\begin{aligned} H^1(S_n) &= \mathbf{Z}/2\mathbf{Z}, \quad n \geq 2, \\ H^2(S_n) &= \mathbf{Z}/2 \oplus \mathbf{Z}/2\mathbf{Z}, \quad n \geq 4 \qquad \text{(Schur)}. \end{aligned}$$

The non-trivial element in $H^1(S_n)$ is the signature homomorphism

$$\epsilon_n : S_n \longrightarrow \{\pm 1\}.$$

The cohomology group $H^2(S_n)$, $n \geq 4$, has a $\mathbf{Z}/2\mathbf{Z}$-basis given by $\epsilon_n \cdot \epsilon_n$ (cup-product) and an element s_n corresponding to the central extension

$$1 \longrightarrow C_2 \longrightarrow \tilde{S}_n \longrightarrow S_n \longrightarrow 1$$

which is characterized by the properties:
1. A transposition in S_n lifts to an element of order 2 in $\tilde{S}_n$.
2. A product of two disjoint transpositions lifts to an element of order 4 in $\tilde{S}_n$.

(The element $\epsilon_n \cdot \epsilon_n$ corresponds to the extension S_n' of S_n obtained by taking the pullback:

$$
\begin{array}{ccc}
S_n' & \longrightarrow & C_4 \\
\downarrow & & \downarrow \\
S_n & \longrightarrow & \{\pm 1\}.
\end{array}
$$

This extension is characterized by the property that a transposition lifts to an element of order 4, while a product of two disjoint transpositions lifts to an element of order 2.)

The image of ϵ_n by the restriction map $H^1(S_n) \longrightarrow H^1(A_n)$ is zero. The cohomology group $H^2(A_n)$ is isomorphic to $\mathbf{Z}/2\mathbf{Z}$ for $n \geq 4$ and is generated by $a_n = \mathrm{Res}(s_n)$. The corresponding central extension of A_n is denoted by $\tilde{A}_n$ (or by $2 \cdot A_n$ in ATLAS' style); it is a subgroup of index 2 of $\tilde{S}_n$.

The cohomology classes ϵ_n and s_n can be given the following topological interpretations: the map $S_n \longrightarrow \mathbf{O}_n(\mathbf{R})$ gives a map of classifying spaces

$$BS_n \longrightarrow B\mathbf{O}_n(\mathbf{R}),$$

and corresponding maps on the cohomology rings. But

$$H^*(B\mathbf{O}_n(\mathbf{R})) = \mathbf{F}_2[w_1, \ldots, w_n],$$

where w_i is the ith Stiefel-Whitney class. The class w_1 corresponds to the element ϵ_n in $H^1(BS_n) = H^1(S_n)$, and w_2 gives s_n.

Exercises:
1. Show that $\tilde{A}_4 \simeq \mathbf{SL}_2(\mathbf{F}_3)$, $\tilde{A}_5 \simeq \mathbf{SL}_2(\mathbf{F}_5)$, and $\tilde{A}_6 \simeq \mathbf{SL}_2(\mathbf{F}_9)$.
2. Let $\hat{S}_n$ be the central extension of S_n by $\{\pm 1\}$ corresponding to the element $s_n + \epsilon_n \cdot \epsilon_n$ of $H^2(S_n)$.
   (a) Show that $\tilde{S}_4 \simeq \mathbf{GL}_2(\mathbf{F}_3)$.
   (b) Show that any outer automorphism of S_6 lifts to an isomorphism of $\hat{S}_6$ onto $\tilde{S}_6$.
   (c) Show that the groups $\hat{S}_n$ and $\tilde{S}_n$ are not isomorphic if $n \geq 4$, $n \neq 6$.

9.2 The quadratic form $\mathrm{Tr}\,(x^2)$

Let E be an étale K-algebra of finite rank n over K; it is a product of separable field extensions of K. There is a dictionary between such algebras and conjugacy classes of homomorphisms $e : G_K \longrightarrow S_n$, which works as follows: given E, let $\phi(E)$ be the set of K-algebra homomorphisms $E \longrightarrow K_s$. The set $\phi(E)$ is of cardinality n, and the natural action of G_K on $\phi(E)$ gives the desired homomorphism $e : G_K \longrightarrow S_n$, after identifying $\phi(E)$ with $\{1, \ldots, n\}$. Conversely, E can be constructed as the twist of the split algebra $K \times \cdots \times K$ by the 1-cocycle $e : G_K \longrightarrow S_n = \mathrm{Aut}\,(K^n)$. The image of e is the Galois group of the smallest extension of K over which the algebra E splits; if E is a field, it is $\mathrm{Gal}(E^{\mathrm{gal}}/K)$.

The function $x \mapsto \mathrm{Tr}\,(x^2)$ defines a non-degenerate quadratic form Q_E of rank n over K. This invariant was studied extensively by 19th century mathematicians such as Jacobi and Hermite.

Theorem 9.2.1 *Let $E/K(T)$ be a finite separable extension of degree n. Let $G \subset S_n$ be the corresponding Galois group. Assume that, for all places v of $K(T)$ not equal to ∞, the order of the inertia group at v is odd. Then Q_E is constant, i.e., comes from a form over K.*

Let Λ be the integral closure of $K[T]$ in E, and denote by $\mathcal{D}$ the different. Using the fact that the inertia groups are of odd order, one can show that $\mathcal{D}^{-1}$ is the square of a fractional ideal,

$$\mathcal{D}^{-1} = \mathcal{A}^2.$$

One checks that Q_E induces a strictly non-degenerate quadratic form over $K[T]$ on the $K[T]$-module $\mathcal{A}$. By th. 9.1.3 the result follows. (One may also show that the second residues of Q_E are 0, and apply Milnor's theorem 9.1.2.)

Example: Let E be a regular Galois extension of $\mathbf{Q}(T)$ with Galois group the Monster, obtained from the rigid family $(2A, 3B, 29A)$, cf. §7.4.7. Since two of the conjugacy classes in this family have odd exponent, the form Q_E comes from $\mathbf{Q}$, and does not depend on T. One can prove that it is hyperbolic.

The following theorem is proved in [Se6].

Theorem 9.2.2 *Let E be an étale K-algebra of rank n and discriminant d associated to a homomorphism $e : G_K \longrightarrow S_n$, and let Q_E denote the trace form of E. Then:*
  1. $w_1(Q_E) = e^*\epsilon_n$,
  2. $w_2(Q_E) = e^*s_n + (2)(d)$.

Suppose that $G = e(G_K)$ is contained in A_n, i.e., that d is a square. By th. 9.2.2, we have $w_1(Q_E) = 0$ and $w_2(Q_E) = e^*a_n$. The element e^*a_n is the obstruction to lifting the homomorphism $G_K \longrightarrow A_n$ to a homomorphism $G_K \longrightarrow \tilde{A}_n$. Hence, we have

Corollary 9.2.3 *The homomorphism $e : G_K \longrightarrow A_n$ lifts to a homomorphism $G_K \longrightarrow \tilde{A}_n$ if and only if the Witt invariant $w_2(Q_E)$ is 0.*

9.3 Application to extensions with Galois group $\tilde{A}_n$

The previous corollary applies when e is surjective: an extension E of K of degree n with Galois group A_n can be embedded in an $\tilde{A}_n$-extension if and only if $w_2(Q_E) = 0$.

This will be used to show the following result:

Theorem 9.3.1 (Mestre, [Me2])   *The group $\tilde{A}_n$ has the Gal_T-property for all n.*

(This was already shown for $n \equiv 0, 1 \pmod 8$, and for some other n, by Vila [Vi].)

The proof constructs a covering $C \longrightarrow \mathbf{P}_1$ of degree n whose Galois closure has Galois group A_n, and which has the following additional properties:

1. There is a point of $\mathbf{P}_1$ whose inverse image is a set $\{a_1, \ldots, a_n\}$, where the a_i are rational and distinct.

2. The non-trivial inertia groups are all generated by cycles of order 3.

Let $E/K(T)$ be the degree n extension corresponding to this covering. Condition 2 implies that the quadratic form Q_E comes from $\mathbf{Q}$, by th. 9.2.1. But by condition 1, there is one rational point where this quadratic form is equivalent to the standard one, $\sum X_i^2$. It is easy to see that this implies that Q_E is equivalent to $\sum X_i^2$ over $\mathbf{Q}(T)$, and hence has trivial Witt invariant; by cor. 9.2.3, we can thus solve the embedding problem for this extension.

Let us say that a property Σ of a polynomial

$$P = X^n + s_1 X^{n-1} + \cdots + s_n$$

is *generally true* if there exists a Zariski open dense subset U of $\mathbf{A}^n$ such that P has the property Σ for all $(s_1, \ldots, s_n)$ in U. Mestre's construction (for n odd) relies on the following:

Proposition 9.3.2 *It is generally true that there exist polynomials Q and R in $\mathbf{Q}[T]$ of degree $n - 1$ with the following properties:*
a) $Q'P - P'Q = R^2$ *(i.e., $(Q/P)' = (R/P)^2$).*
b) P, Q, R *are pairwise relatively prime.*
c) *The zeros $b_1, \ldots, b_{n-1}$ of R in $\bar{\mathbf{Q}}$ are distinct.*
d) *The values of P/Q at the b_i are distinct.*

Sketch of proof: The matrix M with ij-entry $1/(a_i - a_j)$ for $i \neq j$ and 0 for $i = j$ is a skew symmetric matrix of odd dimension. Hence it has zero determinant and there exists a non-zero $(c_1, \ldots, c_n)$ in the kernel of M. We have:

$$\sum_{\substack{j=1 \\ j \neq i}}^{n} \frac{c_j}{a_i - a_j} = 0 \qquad \text{for all } i.$$

Now, let Q, R be the polynomials such that:

$$Q/P = \sum_{i=1}^{n} \frac{-c_i^2}{X - a_i}.$$

$$R/P = \sum_{i=1}^{n} \frac{c_i}{X - a_i}.$$

One has:

$$
\begin{aligned}
(R/P)^2 &= \sum_{i,j=1}^{n} \frac{c_i c_j}{(X - a_i)(X - a_j)} \\
&= \sum_{i=1}^{n} \frac{c_i^2}{(X - a_i)^2} + \sum_{i \neq j} \frac{c_i c_j}{a_i - a_j}\Big(\frac{1}{X - a_i} - \frac{1}{X - a_j}\Big) \\
&= (Q/P)'
\end{aligned}
$$

Moreover, one can check (by looking at a well-chosen example) that it is generally true that M has rank $n-1$ (so that the c_i are essentially unique) and that b), c), d) hold.

If (P, Q, R) are chosen as in prop. 9.3.2, with $P = \prod(X - a_i)$ with $a_i \in \mathbf{Q}$, the a_i being distinct, then the map $\mathbf{P}_1 \longrightarrow \mathbf{P}_1$ given by $X \mapsto T = P(X)/Q(X)$ has degree n, and is ramified at the zeros b_i of R, the ramification groups being generated by 3-cycles. Let $G \subset S_n$ (resp. $\bar{G} \subset S_n$) be the Galois group of the corresponding Galois extension of $\mathbf{Q}(T)$ (resp. of $\bar{\mathbf{Q}}(T)$). The group $\bar{G}$ is transitive and generated by 3-cycles (cf. prop. 4.4.6). By lemma 4.4.4, it is equal to A_n. Hence G is equal to A_n or S_n. However, we have seen that the $\text{Tr}\,(x^2)$ form attached to the extension is the standard form $\sum X_i^2$. In particular, its discriminant is a square. This shows that G is contained in A_n, hence $G = \bar{G} = A_n$, QED.

There is a similar, but more complicated, construction when n is even. Exercise 2 below proves that $\tilde{A}_n$ has the Gal_T property for n even by reducing to the case of odd n.

Remarks:

1. One may also prove Mestre's theorem by showing that the Witt invariant $w_2 \in \text{Br}_2(\mathbf{Q}(T))$ of the trace form has "no poles" (because all the

ramification is odd), and hence is constant, i.e., belongs to $\mathrm{Br}_2(\mathbf{Q})$. Since it is 0 at the base point, it is zero. Another possibility is to prove by a direct construction that the trace form of $E/K(T)$ is constant, cf. exerc. 1.

2. For explicit formulas related to the above constructions, see [Cr], [Schn].

3. For $n = 6, 7$, the Schur multiplier of A_n is cyclic of order 6. The corresponding groups $6 \cdot A_6$ and $6 \cdot A_7$ also have property Gal_T. This has been proved by J-F. Mestre [Me3].

Exercises:

1. Assume P, Q, R are as in prop. 9.3.2 (n odd). Put

$$E = K(X) \text{ and } T = P(X)/Q(X),$$

so that $[E : K(T)] = n$.

a) Let Λ be the integral closure of $K[T]$ in E. Show that $\Lambda = K[X, 1/Q(X)]$ and that the different of Λ over $K[T]$ is the principal ideal generated by $R(X)^2$.

b) Let

$$\mathrm{Tr} : E \longrightarrow K(T)$$

be the trace map. If $f \in K[X]$, show that $\mathrm{Tr}\,(f(X)/R(X)^2)$ belongs to $K[T]$ and that:

$$\deg_T \mathrm{Tr}\,(f(X)/R(X)^2) \leq \sup(0, \deg(f) - 2n + 2).$$

In particular, $\mathrm{Tr}\,(f(X)/R(X)^2)$ belongs to K if $\deg(f) \leq 2n - 2$.

c) Let V be the n-dimensional K-subspace of E spanned by

$$1/R(X), X/R(X), \ldots, X^{n-1}/R(X).$$

One has $E = V \otimes_K K(X)$. Show, using b), that if $v_1, v_2 \in V$, then $\mathrm{Tr}\,(v_1 v_2) \in K$. Conclude that the trace form of $E/K(T)$ comes from K.

2. Let n be even, let $f : \mathbf{P}_1 \longrightarrow \mathbf{P}_1$ be an $n + 1$-covering given by Mestre's construction above, and let C_f be it Galois closure. One has maps

$$C_f \xrightarrow{\;g\;} \mathbf{P}_1 \xrightarrow{\;f\;} \mathbf{P}_1,$$

and $g : C_f \longrightarrow \mathbf{P}_1$ is a regular A_n-covering. Show that this covering lifts to a regular $\tilde{A}_n$-covering of $\mathbf{P}_1$ (hence $\tilde{A}_n$ also has property Gal_T).

Chapter 10

Appendix: the large sieve inequality

10.1 Statement of the theorem

Let N be an integer ≥ 1, and, for each prime p, let ν_p be a real number with $0 < \nu_p \leq 1$. Let A be a subset of $\Lambda = \mathbf{Z}^n$, such that for all primes p,

$$|A_p| \leq \nu_p p^n,$$

where $A_p \subset \Lambda/p\Lambda$ denotes the reduction of A mod p. Given a vector $x = (x_1, \ldots, x_n) \in \mathbf{R}^n$, and $N \in \mathbf{R}$, we denote by $A(x, N)$ the set of points in A which are contained in the cube of side length N centered at x, i.e.,

$$A(x, N) = \{(a_1, \ldots, a_n) \in A \quad | \quad |x_i - a_i| \leq N/2\}.$$

Then:

Theorem 10.1.1 (Large sieve inequality) *For every $D \geq 1$, we have*

$$|A(x, N)| \leq 2^n \sup(N, D^2)^n / L(D),$$

where

$$L(D) = \sum_{\substack{1 \leq d \leq D \\ d \text{ square-free}}} \prod_{p|d} \left(\frac{1 - \nu_p}{\nu_p} \right).$$

Taking $D = N^{\frac{1}{2}}$:

Corollary 10.1.2 $|A(x, N)| \leq (2N)^n / L(N^{\frac{1}{2}}).$

Examples:

1. If $\nu_p = \frac{1}{2}$ for every p, then

$$L(D) = (\sum_{\substack{1 \leq d \leq D \\ d \text{ square}-\text{free}}} 1) \sim \frac{6}{\pi^2} D.$$

Hence $|A(x, N)| << N^{n-\frac{1}{2}}$. This is a typical "large sieve" situation.

2. Assume there is a set S of primes of density > 0, such that $\nu_p = C$ for $p \in S$, with $0 < C < 1$. One may estimate $L(D)$ from below by summing over primes $\leq D$:

$$L(D) \geq 1 + \sum_{\substack{p \leq D \\ p \text{ prime}}} \frac{1 - \nu_p}{\nu_p} \gg \frac{D}{\log D}.$$

Hence $|A(x, N)| \ll N^{n-\frac{1}{2}} \log N$. A more careful estimate of $L(D)$ by summing over all square-free $d \leq D$ allows one to replace the factor $\log N$ by $(\log N)^{\gamma}$, with $\gamma < 1$, under a mild extra condition on S, cf. [Se9], chap. 13.

3. Suppose $n = 1$, and $\nu_p = 1 - \frac{1}{p}$. Then one can show that $L(D) \overset{\smile}{\sim} \log D$, and hence $|A(x, N)| \ll \frac{N}{\log N}$, a weak form of the prime number theorem: however, the method also allows one to conclude that in *any* interval of length N, there are at most $O(\frac{N}{\log N})$ primes. (More precisely, their number is $\leq 2N/\log N$, cf. [MoV].)

Historically, a weaker form of the sieve inequality was discovered first, where the sum giving $L(D)$ was taken over the *primes* $\leq D$; this only gave interesting results in large sieve situations (hence the name "large sieve inequality"). The possibility of using square-free d's was pointed out by Montgomery, [Mo1].

Exercise: Use th. 10.1.1 to show that the number of "twin primes" (primes p such that $p + 2$ is also prime) $\leq N$ is asymptotically $\ll \frac{N}{(\log N)^2}$. Conclude that

$$\sum_{\substack{p \\ \text{twin prime}}} \frac{1}{p} < \infty.$$

Proof of th. 10.1.1: preliminaries

Let us assume without loss of generality that $A = A(x, N)$. Given vectors $a = (a_1, \ldots, a_n)$, $t = (t_1, \ldots, t_n)$ belonging to $\mathbf{R}^n$, put

$$\chi_a(t) = \exp\left(2\pi i \sum_{j=1}^{n} a_j t_j \right).$$

We identify $\Lambda = \mathbf{Z}^N$ with the character group of the torus $T = \mathbf{R}^n/\mathbf{Z}^n$ by $a \mapsto \chi_a$, and associate to $A = A(x, N)$ the function ϕ whose Fourier expansion is:

$$\phi = \sum_{a \in A} \chi_a.$$

The condition on the reduction of A mod p and the fact that A is contained in a cube of side length N give rise to inequalities satisfied by ϕ; combining these will give the sieve inequality.

10.2 A lemma on finite groups

Let C_i for $1 \leq i \leq h$ be finite abelian groups (written additively), $\hat{C}_i$ their character groups, ϕ a function on $C = \prod C_i$. Suppose there are subsets Ω_i of $\hat{C}_i$ with $|\Omega_i| \leq \nu_i |C_i|$, such that the Fourier coefficient of ϕ relative to the character $\chi = (\chi_i) \in \hat{C} = \prod \hat{C}_i$ is 0 outside $\prod \Omega_i$. Let us call $x \in C$ *primitive* if its image in each C_i is $\neq 0$. Then:

Lemma 10.2.1 *We have*:

$$\sum_{\substack{x \in C \\ x \text{ primitive}}} |\phi(x)|^2 \geq |\phi(0)|^2 \prod_i \left(\frac{1 - \nu_i}{\nu_i}\right).$$

We give the proof in the case of a single group C: the general case follows by induction on the number of factors. Write $\phi = \sum c_\chi \chi$, the sum being taken over all characters $\chi \in \Omega$. Then:

$$\sum |c_\chi|^2 = \frac{1}{|C|} \sum_{x \in C} |\phi(x)|^2.$$

Applying the Cauchy-Schwarz inequality, we get

$$|\phi(0)|^2 = |\sum c_\chi \cdot 1|^2 \leq \sum |c_\chi|^2 \sum_{\chi \in \Omega} 1,$$

and hence

$$|\phi(0)|^2 \leq \nu_1 (\sum_{x \neq 0} |\phi(x)|^2 + |\phi(0)|^2).$$

The lemma follows by rearranging terms in this inequality.

10.3 The Davenport-Halberstam theorem

Define a distance on $\mathbf{R}^n$ by $|x| = \sup |x_i|$; this defines a distance on the torus $T = \mathbf{R}^n/\mathbf{Z}^n$, which we also denote by $|\ |$. Let $\delta > 0$; a set of points $\{t_i\}$ in T is called δ-spaced if $|t_i - t_j| \geq \delta$ for all $i \neq j$.

Theorem 10.3.1 (Davenport-Halberstam) *Let* $\phi = \sum c_\lambda \chi_\lambda$ *be a continuous function on* T *whose Fourier coefficients* c_λ *vanish when* λ *is outside some cube* Σ *of size* N. *Let* $t_i \in T$ *be* δ-*spaced points for some* $\delta > 0$. *Then*

$$\sum_i |\phi(t_i)|^2 \leq 2^n \sup(N, \frac{1}{\delta})^n \|\phi\|_2^2,$$

where $\|\phi\|_2$ *is the* L^2-*norm of* ϕ.

If $\delta > 1/2$, there is at most one t_i and the inequality follows from the Cauchy-Schwarz inequality applied to the Fourier expansion of ϕ. Let us now suppose that $\delta \leq \frac{1}{2}$. One constructs an auxiliary function θ on $\mathbf{R}^n$, such that

1. θ is continuous and vanishes outside the cube $|x| < \frac{1}{2}\delta$. This allows us to view θ as a function on T.

2. The Fourier transform of θ has absolute value ≥ 1 on the cube Σ.

3. $\|\theta\|_2^2 \leq 2^n M^n$, where $M = \sup(N, \frac{1}{\delta})$.

Let $\lambda \in \mathbf{R}^n$ be the center of the cube Σ. Then one checks, by an elementary computation, that the function θ defined by

$$\theta(x) = \begin{cases} \chi_\lambda(x) M^n \prod 2 \cos \pi M x_i & \text{if } |x| \leq \frac{1}{2M} \\ \\ 0 & \text{elsewhere.} \end{cases}$$

has the required properties. For each $\lambda \in \Lambda$, let $c_\lambda(\phi)$ be the λ-th Fourier coefficient of ϕ; define similarly $c_\lambda(\theta)$. We have:

$$c_\lambda(\phi) = 0 \text{ if } \lambda \notin \Sigma \quad \text{and} \quad |c_\lambda(\theta)| \geq 1 \text{ if } \lambda \in \Sigma.$$

We may thus define a continuous function g on T whose Fourier coefficients are:

$$c_\lambda(g) = \begin{cases} c_\lambda(\phi)/c_\lambda(\theta) & \text{if } \lambda \in \Sigma \\ \\ 0 & \text{if } \lambda \notin \Sigma \end{cases}$$

Since $c_\lambda(\phi) = c_\lambda(\theta) c_\lambda(g)$ for every $\lambda \in \Lambda$, ϕ is equal to the convolution product $\theta * g$ of θ and g. Therefore:

$$\phi(t_i) = \int_T \theta(t_i - t) g(t) dt = \int_{B_i} \theta(t_i - t) g(t) dt,$$

where B_i is the set of t such that $|t - t_i| < \frac{\delta}{2}$. By the Cauchy-Schwarz inequality:

$$|\phi(t_i)|^2 \leq \|\theta\|_2^2 \int_{B_i} |g(t)|^2 dt \leq 2^n M^n \int_{B_i} |g(t)|^2 dt.$$

Since the t_i are δ-spaced, the B_i are disjoint. Summing over i then gives

$$\sum_i |\phi(t_i)|^2 \le 2^n M^n \|g\|_2^2 \le 2^n M^n \|\phi\|_2^2,$$

because

$$\|g\|_2^2 = \sum \left| \frac{c_\lambda(\phi)}{c_\lambda(\theta)} \right|^2 \le \|\phi\|_2^2.$$

This completes the proof.

Remark: In the case $n = 1$, the factor $2\sup(N, \frac{1}{\delta})$ can be improved to $N + \frac{1}{\delta}$ (Selberg, see e.g. [Mo2]); it is likely that a similar improvement holds for any n.

10.4   Combining the information

Let D be given; the set $\{t_i\}$ of all d-division points of T, where d ranges over positive square-free integers $\le D$, is δ-spaced, for $\delta = 1/D^2$. Applying th. 10.3.1 to $\phi = \sum_{a \in A} \chi_a$, we have

$$\sum_i |\phi(t_i)|^2 \le 2^n \sup(N, D^2)^n |A|. \qquad (10.1)$$

On the other hand, for each $d \le D$ square-free, the kernel $T[d]$ of $d : T \longrightarrow T$ splits as

$$T[d] = \prod_{p|d} T[p]$$

and its character group is $\Lambda/d\Lambda = \prod_{p|d} \Lambda/p\Lambda$. Hypothesis (2) on A allows us to apply lemma 10.2.1 to the restriction of ϕ to $T[d]$. We thus obtain

$$\sum_{\substack{t \in T[d] \\ t \text{ of order } d}} |\phi(t)|^2 \ge |A|^2 \prod_{p|d} \frac{1 - \nu_p}{\nu_p}.$$

Hence, by summing over all square-free $d \le D$, we obtain:

$$\sum_i |\phi(t_i)|^2 \ge |A|^2 L(D). \qquad (10.2)$$

Combining equations 10.1 and 10.2 and cancelling a factor of $|A|$ on both sides gives the large sieve inequality. (The case $|A| = 0$ does not pose any problem.) QED.

Remark: A similar statement holds for a number field K; Λ is replaced by $\mathcal{O}_K \times \cdots \times \mathcal{O}_K$, where $\mathcal{O}_K$ denotes the ring of integers of K; the corresponding torus T is then equipped with a natural action of $\mathcal{O}_K$. The technique of the proof is essentially the same as in the case $K = \mathbf{Q}$, see [Se9], ch. 12.

Exercises:
1. Let p_i $(i \in I)$ be integers ≥ 1 such that $(p_i, p_j) = 1$ if $i \neq j$. Let A be a subset of $\mathbf{Z}^n$ contained in a cube of side length N. Let ν_i be such that the reduction of A mod p_i has at most $\nu_i p_i^n$ elements. Show (by the same method as for th. 10.1.1) that

$$|A| \leq 2^n \sup(N, D^2)^n / L(D),$$

with

$$L(D) = \sum_J \prod_{i \in J} (1 - \nu_i)/\nu_i,$$

where the sum runs through all subsets J of I such that $\prod_{i \in J} p_i \leq D$. (This applies for instance when the p_i's are the squares or the cubes of the prime numbers.)

2. Let H be the set of pairs (x, y) of integers $\neq 0$ such that the Hilbert symbol (x, y) is trivial (i.e., the conic $Z^2 - xX^2 - yY^2 = 0$ has a rational point). Show (by using exerc. 1.) that the number of points of H in a cube of side length N is $<< N^2 / \log N$. (This bound is sharp: this has been proved by C. Hooley [Hoo] and C. R. Guo [Guo].)

For generalizations, see [Se10].

Bibliography

[Ab1]   S. Abhyankar. *Coverings of algebraic curves*, Amer. J. of Math. 79 (1957), 825-856.

[Ab2]   S. Abhyankar. *Galois theory on the line in nonzero characteristic*, Bull. A.M.S., 27 (1992), 68-133.

[AS]   M. Aschbacher and S.D. Smith. *The Classification of Quasi-Thin Groups* (two volumes, 1221 pages), AMS Math. Surveys 111-112, 2004.

[ANVERS]   B.J. Birch and W. Kuyk (edit.). *Modular Functions of One Variable IV*, Lect. Notes in Math. 476, Springer-Verlag, 1975.

[ATLAS]   J.H. Conway, R.T. Curtis, S.P. Norton, R.A. Parker, and R.A. Wilson. *Atlas of finite groups: maximal subgroups and ordinary characters for simple groups.* New York: Clarendon press, 1985.

[Bc]   S. Beckmann. *Ramified primes in the field of moduli of branched coverings of curves*, J. of Algebra 125 (1989), 236-255.

[Be1]   G.V. Belyi. *On Galois extensions of a maximal cyclotomic field*, Izv. Akad. Nauk SSSR 43 (1979), 267-276 (= Math. USSR Izv. 14 (1980), 247-256).

[Be2]   G.V. Belyi. *On extensions of the maximal cyclotomic field having a given classical Galois group*, J. Crelle 341 (1983), 147-156.

[BL]   E. Bayer-Fluckiger and H.W. Lenstra. *Forms in odd degree extensions and self-dual normal bases*, Amer. J. of Math. 112 (1990), 359-373.

[Bo]   E. Bombieri. *On exponential sums in finite fields II*, Invent.
              Math. 47 (1978), 29-39.

[Bor]   A. Borel. *Introduction aux groupes arithmétiques*, Paris, Her-
              mann, 1969.

[Coh]   S.D. Cohen. *The distribution of Galois groups and Hilbert's
              irreducibility theorem*, Proc. London Math. Soc.(3) 43 (1981),
              227-250.

[CT]   J-L. Colliot-Thélène. Letter to T. Ekedahl, 9/21/1988, unpub-
              lished.

[CTS1]   J-L. Colliot-Thélène and J-J. Sansuc. *Principal homogeneous
              spaces and flasque tori: applications*, J. of Algebra 106 (1987),
              148-205.

[CTS2]   J-L. Colliot-Thélène and J-J. Sansuc. *La descente sur les
              variétés rationnelles, II*, Duke Math. J. 54 (1987), 375-492.

[Cr]   T. Crespo. *Explicit construction of $\tilde{A}_n$-type fields*, J. of Algebra
              127 (1989), 452-461.

[CR]   C.W. Curtis and I. Reiner. *Representation Theory of Finite
              Groups and Associative Algebras*, Inters. Publ., 1962.

[Da]   H. Davenport. *Multiplicative Number Theory* (revised by H.L.
              Montgomery), Springer-Verlag, 1980.

[DD]   R. and A.Douady. *Algèbre et théories galoisiennes; 2/ théories
              galoisiennes*, CEDIC/F.Nathan, Paris, 1979.

[De1]   P. Deligne. *Le support du caractère d'une représentation su-
              percuspidale*, C.R. Acad. Sci. Paris 283 (1976), 155-157.

[De2]   P. Deligne. *Le groupe fondamental du complément d'une
              courbe plane n'ayant que des points doubles ordinaires est
              abélien*, Sém. Bourbaki 1979/80, no. 543.

[Ek]   T. Ekedahl. *An effective version of Hilbert's irreducibility the-
              orem*, preprint, Stockholm, 1987. (cf. sém. Théorie des Nom-
              bres, Paris 1988-1989, Birkhäuser, 1990, 241-248).

[F]   R.Fricke. *Lehrbuch der Algebra. III. Algebraische Zahlen*,
              Braunschweig, 1928.

[Fo]   O. Forster. *Lectures on Riemann surfaces*, Springer-Verlag,
              1981.

[Fr1]   M.D. Fried. *Fields of definition of function fields and Hurwitz families - Groups as Galois groups*, Comm. Alg. 5 (1977), 17-82.

[Fr2]   M.D. Fried. *Rigidity and applications of the classification of simple groups to monodromy*, preprint.

[Ga]   P.X. Gallagher. *The large sieve and probabilistic Galois theory*, Proc. Symp. Pure Math. 24, 91-101, A.M.S., 1973.

[GH]   P. Griffiths and J. Harris. *Principles of Algebraic Geometry*, John Wiley and Sons, New York, 1978.

[GR]   H. Grauert and R. Remmert. *Komplexe Räume*, Math. Ann. 136 (1958), 245-318.

[Guo]   C.R. Guo. *On solvability of ternary quadratic forms*, Proc. London Math. Soc. (3) 70 (1995), 241-263.

[Harb]   D. Harbater. *Abhyankar's conjecture on Galois groups over curves*, Invent. Math. 117 (1994), 1-25.

[Ha]   R. Hartshorne. *Algebraic Geometry*, Springer-Verlag, 1977.

[He]   E. Hecke. *Die eindeutige Bestimmung der Modulfunktionen q-ter Stufe durch algebraische Eigenschaften*, Math. Ann. 111 (1935), 293-301 (= Math. Werke, 568-576).

[Hi]   D. Hilbert. *Ueber die Irreduzibilität ganzer rationaler Funktionen mit ganzzahligen Koeffizienten*, J. Crelle 110 (1892), 104-129 (= Ges. Abh. II, 264-286).

[Ho]   G. Hoyden-Siedersleben. *Realisierung der Jankogruppen J_1 und J_2 als Galoisgruppen über* **Q**, J. of Algebra 97 (1985), 14-22.

[Hoo]   C. Hooley, *On ternary quadratic forms which represent zero*, Glasgow Math. J. 35 (1993), 13-23.

[Hu]   B. Huppert. *Endliche Gruppen I*, Springer-Verlag, 1967.

[Hunt]   D.C. Hunt. *Rational rigidity and the sporadic groups*, J. of Algebra 99 (1986), 577-592.

[Is]   V.V. Ishkhanov. *On the semi-direct embedding problem with nilpotent kernel*, Izv. Akad. Nauk SSSR 40 (1976), 3-25, (=Math. USSR Izv. 10 (1976),1-23).

[J1]   C. Jordan. *Traité des substitutions et des équations algébriques*, Gauthier-Villars, Paris, 1870; 2ème édition, Paris, 1957.

[J2]   C. Jordan. *Recherches sur les substitutions*, J. Liouville 17 (1872), 351-367 (= Oeuvres, I, no. 52).

[J3]   C. Jordan. *Sur la limite de transitivité des groupes non alternés*, Bull. Soc. Math. France 1 (1873), 40-71 (= Oeuvres, I, no. 57).

[Jou]   J-P. Jouanolou. *Théorèmes de Bertini et applications*, Birkhäuser, 1983.

[JY]   C.U. Jensen and N. Yui. *Quaternion extensions*, Algebraic Geometry and Commutative Algebra, in honor of M. Nagata (1987), 155-182.

[Ki]   R. Kiehl. *Der Endlichkeitsatz für eigentliche Abbildungen in der nichtarchimedischen Funktionentheorie*, Invent. Math. 2 (1967), 191-214.

[Kö]   U. Köpf. *Über eigentliche Familien algebraischer Varietäten über affinoiden Räumen*, Schriftenreihe Univ. Münster, 2. serie Heft 7, 1974.

[KN]   W. Krull and J. Neukirch. *Die Struktur der absoluten Galoisgruppe über dem Körper* $\mathbf{R}(t)$, Math. Ann. 193 (1971), 197-209.

[Ku]   W. Kuyk. *Extensions de corps hilbertiens*, J. of Algebra 14 (1970), 112-124.

[L]   S. Lang. *Fundamentals of Diophantine Geometry*, Springer-Verlag, New York, 1983.

[Le]   H.W. Lenstra. *Rational functions invariant under a finite abelian group*, Invent. Math. 25 (1974), 299-325.

[LM]   M.E. LaMacchia. *Polynomials with Galois group* $\mathbf{PSL}(2,7)$, Commun. Alg. 8 (1980), 983-992.

[Mae]   T. Maeda. *Noether's problem for* A_5, J. of Algebra 125 (1989), 418-430.

[Ma1]   B.H. Matzat. *Zwei Aspekte konstruktiver Galoistheorie*, J. of Algebra 96 (1985), 499-531.

[Ma2]   B.H. Matzat. *Topologische Automorphismen in der konstruktiver Galoistheorie*, J. Crelle 371 (1986), 16-45.

[Ma3]   B.H. Matzat. *Konstruktive Galoistheorie*, Lect. Notes in Math. 1284, Springer-Verlag, 1987.

[Ma4]   B.H. Matzat. *Konstruktion von Zahlkörpern mit der Galoisgruppe M_{11} über* $\mathbf{Q}(\sqrt{-11})$, Man. Math. 27 (1979), 103-111.

[Me1]   J-F. Mestre. *Courbes hyperelliptiques à multiplications réelles*, C.R.Acad.Sci.Paris 307 (1988), 721-724.

[Me2]   J-F. Mestre. *Extensions régulières de* $\mathbf{Q}(T)$ *de groupe de Galois* $\tilde{A}_n$, J. of Algebra 131 (1990), 483-495.

[Me3]   J-F. Mestre. *Extensions* $\mathbf{Q}$-*régulières de* $\mathbf{Q}(T)$ *de groupe de Galois 6.A6 et 6.A7*, Israel J. Math. 107 (1998), 333-341.

[Ml1]   G. Malle. *Exceptional groups of Lie type as Galois groups*, J. Crelle 392 (1988), 70-109.

[Ml2]   G. Malle. *Genus zero translates of three point ramified Galois extensions*, Man. Math. 71 (1991), 97-111.

[MM]   G. Malle and B.H. Matzat. *Inverse Galois Theory*, Springer-Verlag, 1999.

[Mo1]   H.L. Montgomery. *A note on the large sieve*, J. London Math. Soc. 43 (1968), 93-98.

[Mo2]   H.L. Montgomery. *The analytic principle of the large sieve*, Bull. A.M.S. 84 (1978), 547-567.

[MoV]   H.L. Mongomery and R.C. Vaughan. *The large sieve*, Mathematica 20 (1973), 119-134.

[Ne]   J. Neukirch. *On solvable number fields*, Invent. Math. 53 (1979), 135-164.

[Noe]   E. Noether. *Gleichungen mit vorgeschriebener Gruppe*, Math. Ann. 78 (1918), 221-229 (=Ges. Abh. no. 42).

[NSW]   J. Neukirch, A. Schmidt and K. Wingberg. *Cohomology of Number Fields*, Springer-Verlag, 2000.

[Oe]   J. Oesterlé. *Nombres de Tamagawa et groupes unipotents en caractéristique p*, Invent. Math 78 (1984), 13-88.

[Os]   H. Osada. *The Galois groups of the polynomials $x^n + ax^l + b$*,
            J. Number Theory 25 (1987), 230-238.

[Pa]   H. Pahlings. *Some sporadic groups as Galois groups*, Rend.
            Sem. Mat. Univ. Padova 79 (1988), 97-107; II, ibid. 82 (1989),
            163-171; correction, ibid. 85 (1991), 309-310.

[Pr]   B. Pryzwara. *Die Operation der Hurwitzschen Zopfgruppe auf
            den Erzeugendensystemklassen endlicher Gruppen*, Diplomar-
            beit, Karlsruhe, 1988.

[Ra1]   M. Raynaud. *Anneaux locaux henséliens*, Lect. Notes in Math.
            169, Springer-Verlag, 1970.

[Ra2]   M. Raynaud. *Revêtements de la droite affine en caractéristique
            $p > 0$ et conjecture d'Abhyankar*, Invent. Math. 116 (1994),
            425-462.

[Re]   H. Reichardt. *Konstruktion von Zahlkörpern mit gegebener
            Galoisgruppe von Primzahlpotenzordnung*, J. Crelle 177
            (1937), 1-5.

[RR]   M.S. Raghunathan and A. Ramanathan. *Principal bundles on
            the affine line*, Proc. Indian Acad. Sci. (Math. Sci.) 93 (1984),
            137-145.

[Sa1]   D. Saltman. *Generic Galois extensions and problems in field
            theory*, Adv. Math. 43 (1982), 250-283.

[Sa2]   D. Saltman. *Noether's problem over an algebraically closed
            field*, Invent. Math. 77 (1984), 71-84.

[Sch]   W. Scharlau. *Quadratic and Hermitian Forms*, Springer-
            Verlag, 1985.

[Schm]   W.M. Schmidt. *Integer points on hypersurfaces*, Monat. für
            Math. 102 (1986), pp. 27-58.

[Schn]   L. Schneps. *Explicit construction of extensions of $K(t)$ of Ga-
            lois group $\tilde{A}_n$ for n odd*, J. of Algebra 146 (1992), 117-123.

[Se1]   J-P. Serre. *Cohomologie Galoisienne*, Lect. Notes in Math. 5,
            Springer-Verlag, 1994 (5th revised edition); English transla-
            tion: *Galois Cohomology*, SMM, Springer-Verlag, 2002.

[Se2]   J-P. Serre. *Local Fields*, GTM 67, Springer-Verlag, 1979.

[Se3]   J-P. Serre. *Algebraic Groups and Class Fields*, GTM 117,
             Springer-Verlag, 1988.

[Se4]   J-P. Serre. *Géométrie algébrique et géométrie analytique*, Ann.
             Inst. Fourier 6 (1956), 1-42 (=C.P. no. 32).

[Se5]   J-P. Serre. *Extensions icosaédriques*, Sém. Th. des Nombres,
             Bordeaux, 1979-80, exposé 19 (=C.P. no. 123).

[Se6]   J-P. Serre. *L'invariant de Witt de la forme* $\mathrm{Tr}\,(x^2)$, Comm.
             Math. Helv. 59 (1984), 651-679. (=C.P. no. 131).

[Se7]   J-P. Serre. *Modular forms of weight one and Galois represen-
             tations*, Algebraic Number Fields (A. Fröhlich edit.) Acad.
             Press, 1977, 193-268. (=C.P. no. 110).

[Se8]   J-P. Serre. *Groupes de Galois sur* **Q**, Sém. Bourbaki 1987-
             1988, no. 689. (=C.P. no. 147).

[Se9]   J-P. Serre. *Lectures on the Mordell-Weil theorem*, translated
             and edited by M. Brown from notes by M. Waldschmidt,
             Vieweg-Verlag, 1989.

[Se10]   J-P. Serre. *Spécialisation des éléments de* $\mathrm{Br}_2(\mathbf{Q}(T_1,\ldots,T_n))$,
             C.R. Acad. Sci. Paris 311 (1990), 397-402. (=C.P. no. 150).

[Se11]   J-P. Serre. *Revêtements des courbes algébriques*, Sém. Bour-
             baki, 1991-1992, no. 749. (=C.P. no. 157).

[Se12]   J-P. Serre. *On a theorem of Jordan*, Math. Medley 29 (2002),
             3-18; Bull. AMS 40 (2003), 429-440.

[Sel]   E.S. Selmer. *On the irreducibility of certain trinomials*, Math.
             Scand. 4 (1956), 287-302.

[SGA1]   A. Grothendieck. *Revêtement étales et groupe fondamental*,
             Lect. Notes in Math. 224, Springer-Verlag, 1971.

[Sha1]   I.R. Shafarevich. *Construction of fields of algebraic numbers
             with given solvable Galois group*, Izv. Akad. Nauk SSSR 18
             (1954), 525-578 (=Amer. Math. Transl. 4 (1956), 185-237 =
             C.P., 139-191).

[Sha2]   I.R. Shafarevich. *The embedding problem for split extensions*,
             Dokl. Akad. Nauk SSSR 120 (1958), 1217-1219 (=C.P., 205-
             207).

[Sha3]   I.R. Shafarevich. *On the factors of a descending central series* (in Russian), Mat.Zam. 45 (1989), 114-117.

[Shih1]   K-y. Shih. *On the construction of Galois extensions of function fields and number fields*, Math. Ann. 207 (1974), 99-120.

[Shih2]   K-y. Shih. *p-division points on certain elliptic curves*, Comp. Math. 36 (1978), 113-129.

[Shim]   G. Shimura. *Introduction to the arithmetic theory of automorphic functions*, Publ. Math.Soc.Japan 11, Iwanami Shoten and Princeton University Press, 1971.

[Sie]   C.L. Siegel. *Topics in complex function theory, vol. 2*, Wiley-Interscience, New York, 1971.

[Sw1]   R. Swan. *Invariant rational functions and a problem of Steenrod*, Invent. Math. 7 (1969), 148-158.

[Sw2]   R. Swan. *Noether's problem in Galois theory*, Emmy Noether in Bryn Mawr (J.D. Sally and B. Srinivasan edit.), 21-40, Springer-Verlag, 1983.

[Th1]   J.G. Thompson. *A non-duality theorem for finite groups*, J. of Algebra 14 (1970), 1-4.

[Th2]   J.G. Thompson. *Some finite groups which appear as* $\mathrm{Gal}(L/K)$, *where* $K \subseteq \mathbf{Q}(\mu_n)$, J. of Algebra 89 (1984), 437-499.

[Th3]   J.G. Thompson. $\mathbf{PSL}_3$ *and Galois groups over* $\mathbf{Q}$, Proc. Rutgers group theory year 1983-1984, Cambridge Univ. Press, 1984, 309-319.

[To]   D. Toledo. *Projective varieties with non-residually finite fundamental group*, Publ. Math. IHES 77 (1993), 103-119.

[Vi]   N. Vila. *On central extensions of* A_n *as Galois group over* $\mathbf{Q}$, Arch. Math. 44 (1985), 424-437.

[Vo1]   V.E. Voskresenskii. *Birational properties of linear algebraic groups*, Izv. Akad. Nauk SSSR 34 (1970), 3-19 (= Math. USSR Izv. 4 (1970), 1-17).

[Vo2]   V.E. Voskresenskii. *On the question of the structure of the subfields of invariants of a cyclic group of automorphisms of the field* $\mathbf{Q}(x_1, \ldots, x_n)$, Izv. Akad. Nauk SSSR 34 (1970), 366-375 (=Math. USSR Izv. 4 (1970), 371-380).

[Wa]   S. Wang. *A counterexample to Grunwald's theorem*, Ann. Math. 49 (1948), 1008-1009.

[We]   A. Weil. *Adeles and algebraic groups*. Notes by M. Demazure and T. Ono, Birkhäuser-Boston, 1982.

[Wi]   H. Wielandt. *Finite Permutation Groups*, Acad. Press, New York, 1964.

[Z]   O. Zariski. *Pencils on an algebraic variety and a new proof of a theorem of Bertini*, Trans. A.M.S. 50 (1941), 48-70 (=C.P., I, 154-176).

Index

T - #0042 - 160425 - C0 - 229/152/8 [10] - CB - 9781568814124 - Gloss Lamination